Undergraduate Texts in Mathematics

Undergraduate Texts in Mathematics

Apostol: Introduction to Analytic Number Theory.
1976. xii, 338 pages. 24 illus.

Armstrong: Basic Topology.
1983. xii, 260 pages. 132 illus.

Bak/Newman: Complex Analysis.
1982. x, 224 pages. 69 illus.

Banchoff/Wermer: Linear Algebra Through Geometry.
1983. x, 257 pages. 81 illus.

Childs: A Concrete Introduction to Higher Algebra.
1979. xiv, 338 pages. 8 illus.

Chung: Elementary Probability Theory with Stochastic Processes.
1975. xvi, 325 pages. 36 illus.

Croom: Basic Concepts of Algebraic Topology.
1978. x, 177 pages. 46 illus.

Ebbinghaus/Flum/Thomas Mathematical Logic.
1984. xii, 216 pages. 1 illus.

Fischer: Intermediate Real Analysis.
1983. xiv, 770 pages. 100 illus.

Fleming: Functions of Several Variables. Second edition.
1977. xi, 411 pages. 96 illus.

Foulds: Optimization Techniques: An Introduction.
1981. xii, 502 pages. 72 illus.

Franklin: Methods of Mathematical Economics. Linear and Nonlinear Programming. Fixed-Point Theorems.
1980. x, 297 pages. 38 illus.

Halmos: Finite-Dimensional Vector Spaces. Second edition.
1974. viii, 200 pages.

Halmos: Naive Set Theory.
1974, vii, 104 pages.

Iooss/Joseph: Elementary Stability and Bifurcation Theory.
1980. xv, 286 pages. 47 illus.

Jänich: Topology
1984. ix, 180 pages (approx.). 180 illus.

Kemeny/Snell: Finite Markov Chains.
1976. ix, 224 pages. 11 illus.

Lang: Undergraduate Analysis
1983. xiii, 545 pages. 52 illus.

Lax/Burstein/Lax: Calculus with Applications and Computing, Volume 1. Corrected Second Printing.
1984. xi, 513 pages. 170 illus.

LeCuyer: College Mathematics with A Programming Language.
1978. xii, 420 pages. 144 illus.

Macki/Strauss: Introduction to Optimal Control Theory.
1981. xiii, 168 pages. 68 illus.

Malitz: Introduction to Mathematical Logic: Set Theory - Computable Functions - Model Theory.
1979. xii, 198 pages. 2 illus.

Martin: The Foundations of Geometry and the Non-Euclidean Plane.
1975. xvi, 509 pages. 263 illus.

Martin: Transformation Geometry: An Introduction to Symmetry.
1982. xii, 237 pages. 209 illus.

continued after Index

H.-D. Ebbinghaus
J. Flum
W. Thomas

Mathematical Logic

Springer-Verlag
New York Berlin Heidelberg Tokyo

H.-D. Ebbinghaus
J. Flum
Mathematisches Institut
Universität Freiburg
Albertstrasse 23b
7800 Freiburg
Federal Republic of Germany

W. Thomas
Lehrstuhl für Informatik II
RWTH Aachen
Büchel 29–31
5100 Aachen
Federal Republic of Germany

Translated from *Einführung in die mathematische Logik*, published by Wissenschaftliche Buchgesellschaft, Darmstadt, by Ann S. Ferebee, Kohlweg 12, D-6240 Königstein 4, Federal Republic of Germany.

AMS Subject Classification (1980): 03–01

Library of Congress Cataloging in Publication Data
Ebbinghaus, Heinz-Dieter, 1939–
Mathematical logic.
(Undergraduate texts in mathematics)
Translation of: Einführung in die mathematische Logik.
Bibliography: p.
Includes index.
1. Logic, Symbolic and mathematical. I. Flum, Jörg.
II. Thomas, Wolfgang. III. Title. IV. Series.
QA9.E2213 1984 511.3 83-20060

With 1 Illustration

Typeset by Composition House Ltd., Salisbury, England.
Printed and bound by R. R. Donnelley & Sons, Harrisonburg, Virginia.
Printed in the United States of America.

9 8 7 6 5 4 3 2 1

ISBN 0-387-90895-1 Springer-Verlag New York Berlin Heidelberg Tokyo
ISBN 3-540-90895-1 Springer-Verlag Berlin Heidelberg New York Tokyo

Preface

Some of the central questions of mathematical logic are: What is a mathematical proof? How can proofs be justified? Are there limitations to provability? To what extent can machines carry out mathematical proofs?

Only in this century has there been success in obtaining substantial and satisfactory answers, the most pleasing of which is given by Gödel's completeness theorem: It is possible to exhibit (in the framework of first-order languages) a simple list of inference rules which suffices to carry out all mathematical proofs. "Negative" results, however, appear in Gödel's incompleteness theorems. They show, for example, that it is impossible to prove all true statements of arithmetic, and thus they reveal principal limitations of the axiomatic method.

This book begins with an introduction to first-order logic and a proof of Gödel's completeness theorem. There follows a short digression into model theory which shows that first-order languages have some deficiencies in expressive power. For example, they do not allow the formulation of an adequate axiom system for arithmetic or analysis. On the other hand, this difficulty can be overcome—even in the framework of first-order logic—by developing mathematics in set-theoretic terms. We explain the prerequisites from set theory that are necessary for this purpose and then treat the subtle relation between logic and set theory in a thorough manner.

Gödel's incompleteness theorems are presented in connection with several related results (such as Trahtenbrot's theorem) which all exemplify the limitations of machine oriented proof methods. The notions of computability theory that are relevant to this discussion are given in detail. The concept of computability is made precise by means of a simple programming language.

The development of mathematics in the framework of first-order logic (as indicated above) makes use of set-theoretic notions to an extent far beyond that of mathematical practice. As an alternative one can consider logical systems with more expressive power. We introduce some of these systems, such as second-order and infinitary logics. In each of these cases we point out deficiencies contrasting first-order logic. Finally, this empirical fact is confirmed by Lindström's theorems, which show that there is no logical system that extends first-order logic and at the same time shares all its advantages.

The book does not require special mathematical knowledge; however, it presupposes an acquaintance with mathematical reasoning as acquired, for example, in the first year of a mathematics or computer science curriculum. Exercises enable the reader to test and deepen his understanding of the text. The references in the bibliography point out essays of historical importance, further investigations, and related fields.

The original edition of the book appeared in 1978 under the title "Einführung in die mathematische Logik." Some sections have been revised for the present translation; furthermore, some exercises have been added. We thank Dr. J. Ward for his assistance in preparing the final English text. Further thanks go to Springer-Verlag for their friendly cooperation.

Freiburg and Aachen
November 1983

H.-D. Ebbinghaus
J. Flum
W. Thomas

Contents

PART A

CHAPTER I
Introduction

Towards the end of the nineteenth century mathematical logic evolved into a subject of its own. It was the works of Boole, Frege, Russell, and Hilbert, among others,[1] that contributed to its rapid development. Various elements of the subject can already be found in traditional logic, for example, in the works of Aristotle or Leibniz. However, while traditional logic can be considered as part of philosophy, mathematical logic is more closely related to mathematics. Some aspects of this relation are:

(1) *Motivation and Goals.* Investigations in mathematical logic arose mainly from questions concerning the foundations of mathematics. For example, Frege intended to base mathematics on logical and set-theoretical principles. Russell tried to eliminate contradictions that arose in Frege's system. Hilbert's goal was to show that "the generally accepted methods of mathematics taken as a whole do not lead to a contradiction" (this is known as Hilbert's program).

(2) *Methods.* In mathematical logic the methods used are primarily *mathematical.* This is exemplified by the way in which new concepts are formed, definitions are given, and arguments are conducted.

(3) *Applications in Mathematics.* The methods and results obtained in mathematical logic are not only useful for treating foundational problems; they also increase the stock of tools available in mathematics itself. There are applications in many areas of mathematics, such as algebra and topology.

[1] Aristotle (384–322 B.C.), G. W. Leibniz (1646–1716), G. Boole (1815–1864), G. Frege (1848–1925), D. Hilbert (1862–1943), B. Russell (1872–1970).

However, these mathematical features do not result in mathematical logic being of interest solely to mathematicians. For example, the mathematical approach leads to a clarification of concepts and problems that also are of importance in traditional logic and in other fields, such as epistemology or the philosophy of science. In this sense the restriction to mathematical methods turns out to be very fruitful.

In mathematical logic, as in traditional logic, *deductions* and *proofs* are central objects of investigation. However, it is the methods of deduction and the types of argument as used in *mathematical* proofs which are considered in mathematical logic (cf. (1)). In the investigations themselves, mathematical methods are applied (cf. (2)). This close relationship between the subject and the method of investigation, particularly in the discussion of foundational problems, may create the impression that we are in danger of becoming trapped in a vicious circle. We shall not be able to discuss this problem in detail until Chapter VII, and we ask the reader who is concerned about it to bear with us until then.

§1. An Example from Group Theory

In this and the next section we present two simple mathematical proofs. They serve as illustrations of some of the methods of proof as used by mathematicians. Guided by these examples we raise some questions which lead us to the main topics of the book.

We begin with the proof of a theorem from group theory. We therefore require the *axioms of group theory*, which we now state. We use $\circ$ to denote the group multiplication and e to denote the identity element. The axioms may then be formulated as follows:

(G1) For all x, y, z: $(x \circ y) \circ z = x \circ (y \circ z)$.
(G2) For all x: $x \circ e = x$.
(G3) For every x there is a y such that $x \circ y = e$.

A *group* is a triple $(G, \circ^G, e^G)$ which satisfies (G1), (G2), and (G3). Here G is a set, e^G is an element of G, and $\circ^G$ is a binary function on G, i.e., a function defined on all pairs of elements from G, the values of which are also elements of G. The variables x, y, z range over elements of G, $\circ$ refers to $\circ^G$, and e refers to e^G.

As an example of a group we mention the additive group of reals $(\mathbb{R}, +, 0)$, where $\mathbb{R}$ is the set of real numbers, $+$ is the usual addition, and 0 is the real number zero. On the other hand, $(\mathbb{R}, \cdot, 1)$ is not a group (where $\cdot$ is the usual multiplication). For example, the real number 0 violates axiom (G3): there is no real number r such that $0 \cdot r = 1$.

We call triples such as $(\mathbb{R}, +, 0)$ or $(\mathbb{R}, \cdot, 1)$ *structures*. In Chapter III we shall give an exact definition of the notion of structure.

Now we prove the following simple theorem from group theory:

1.1 Theorem (Existence of a Left Inverse). *For every x there is a y such that $y \circ x = e$.*

PROOF. Let x be chosen arbitrarily. From (G3) we know that, for a suitable y,

$$x \circ y = e. \tag{1}$$

Again from (G3) we get, for this y, an element z such that

$$y \circ z = e. \tag{2}$$

We can now argue as follows:

$$\begin{aligned} y \circ x &= (y \circ x) \circ e && \text{(by (G2))} \\ &= (y \circ x) \circ (y \circ z) && \text{(from (2))} \\ &= y \circ (x \circ (y \circ z)) && \text{(by (G1))} \\ &= y \circ ((x \circ y) \circ z) && \text{(by (G1))} \\ &= y \circ (e \circ z) && \text{(from (1))} \\ &= (y \circ e) \circ z && \text{(by (G1))} \\ &= y \circ z && \text{(by (G2))} \\ &= e && \text{(from (2)).} \end{aligned}$$

Since x was arbitrary, we conclude that for every x there is a y such that $y \circ x = e$. □

The proof shows that in every structure where (G1), (G2), and (G3) are satisfied, i.e., in every group, the theorem on the existence of a left inverse holds. A mathematician would also describe this situation by saying that the theorem on the existence of a left inverse *follows from*, or *is a consequence of* the axioms of group theory.

§2. An Example from the Theory of Equivalence Relations

The theory of equivalence relations is based on the following three axioms (xRy is to be read "x is equivalent to y"):

(E1) For all x: xRx.
(E2) For all x and y: If xRy, then yRx.
(E3) For all x, y, z: If xRy and yRz, then xRz.

Let A be a nonempty set, and let R^A be a binary relation on A, i.e., $R^A \subset A \times A$. For $(a, b) \in R^A$ we also write aR^Ab. The pair (A, R^A) is another example of a structure. We call R^A an *equivalence relation on* A, and the structure (A, R^A) an *equivalence structure* if (E1), (E2), and (E3) are satisfied. For example, $(\mathbb{Z}, R_5)$ is an equivalence structure, where $\mathbb{Z}$ is the set of integers and

$$R_5 = \{(a, b) \mid a, b \in \mathbb{Z} \text{ and } b - a \text{ is divisible by } 5\}.$$

On the other hand, the binary relation R_{rp} on $\mathbb{Z}$, which holds between two integers if they are relatively prime, is not an equivalence relation over $\mathbb{Z}$. For example, 5 and 7 are relatively prime, and 7 and 15 are relatively prime, but 5 and 15 are not relatively prime; thus (E3) does not hold for R_{rp}.

We now prove a simple theorem about equivalence relations.

2.1 Theorem. *If x and y are both equivalent to a third element, they are equivalent to the same elements. More formally, for all x and y, if there is a u such that xRu and yRu, then for all z, xRz if and only if yRz.*

PROOF. Let x and y be given arbitrarily; suppose that for some u

$$xRu \quad \text{and} \quad yRu. \tag{1}$$

From (E2) we then obtain

$$uRx \quad \text{and} \quad uRy. \tag{2}$$

From xRu and uRy we deduce, using (E3),

$$xRy, \tag{3}$$

and from yRu and uRx we likewise get (using (E3))

$$yRx. \tag{4}$$

Now let z be chosen arbitrarily. If

$$xRz \tag{5}$$

then, using (E3), we obtain from (4) and (5)

$$yRz.$$

On the other hand, if

$$yRz \tag{6}$$

then, using (E3), we get from (3) and (6)

$$xRz.$$

Thus the claim is proved for all z. □

As in the previous example, this proof shows that every structure (of the form (A, R^A)) which satisfies the axioms (E1), (E2), and (E3), also satisfies Theorem 2.1, i.e., that 2.1 follows from (E1), (E2), and (E3).

§3. A Preliminary Analysis

We sketch some aspects which the two examples just given have in common. In each case one starts from a system Φ of propositions which is taken to be a *system of axioms* for the theory in question (group theory, theory of equivalence relations). The mathematician is interested in finding the propositions which *follow* from Φ, where a proposition ψ is said to follow from Φ if ψ holds in every structure which satisfies all propositions in Φ. A *proof* of ψ from a system Φ of axioms shows that ψ follows from Φ.

When we think about the scope of methods of mathematical proof, we are led to ask about the converse:

(*) Is every proposition ψ which follows from Φ also provable from Φ?

For example, is every proposition which holds in all groups also provable from the group axioms (G1), (G2), and (G3)?

The material developed in Chapters II through V and in Chapter VII yields an essentially positive answer to (*). Clearly it is necessary to make the concepts "proposition", "follows from", and "provable", which occur in (*), more precise. We sketch briefly how we shall do this.

(1) *The Concept "Proposition"*. Normally the mathematician uses his everyday language (e.g., English or German) to formulate his propositions. But since sentences in everyday language are not, in general, completely unambiguous in their meaning and structure, we cannot specify them by precise definitions. For this reason we shall introduce a *formal language L* which reflects features of mathematical statements. Like programming languages used today, L will be formed according to fixed rules: Starting with a set of symbols (an "alphabet"), we obtain so-called *formulas* as finite symbol strings built up in a standard way. These formulas correspond to propositions expressed in everyday language. For example, the symbols of L will include $\forall$ (to be read "for all"), $\wedge$ ("and"), $\rightarrow$ ("if ... then"), $\equiv$ ("equal"), and variables like x, y, and z. Formulas of L will be expressions like

$$\forall x\, x \equiv x, \qquad x \equiv y, \qquad x \equiv z,$$

and

$$\forall x\, \forall y\, \forall z((x \equiv y \wedge y \equiv z) \rightarrow x \equiv z).$$

Although the expressive power of L may at first appear to be limited, we shall later see that many mathematical propositions can be formulated in L. We shall even see that L is in principle sufficient for all of mathematics. The definition of L will be given in Chapter II.

(2) *The Concept "Follows From" (the Consequence Relation).* Axioms (G1)–(G3) of group theory obtain a meaning when interpreted in structures of the form $(G, \circ^G, e^G)$. In an analogous way we can define the general notion of an L-formula holding in a structure. This enables us (in Chapter III) to define the consequence relation: ψ *follows from* (is a consequence of) Φ if and only if ψ holds in every structure where all formulas of Φ hold.

(3) *The Concept "Proof".* A mathematical proof of a proposition ψ from a system Φ of axioms consists of a series of *inferences* which proceeds from axioms of Φ or propositions that have already been proved to new propositions, and which finally ends with ψ. At each step of a proof the mathematician writes something like "From ... and ____ one obtains directly that ~~~~", and he expects it to be clear to anyone that the validity of ... and of ____ entails the validity of ~~~~.

An analysis of examples shows that the grounds for accepting such inferences are often closely related to the meaning of *connectives*, such as "and", "or", or "if-then", and *quantifiers*, "for all" or "there exists", which occur there. For example, this is the case in the first step of the proof of 1.1, where we deduce from "*for all* x there is a y such that $x \circ y = e$" that for the given x there is a y such that $x \circ y = e$. Or consider the step from (1) and (2) to (3) in the proof of 2.1, where from the proposition "xRu *and* yRu" we infer the left member of the conjunction, "xRu", and from "uRx *and* uRy" we infer the right member, "uRy", and then using (E3) we conclude (3).

The formal character of the language L makes it possible to represent these inferences as formal operations on symbol strings (the L-formulas). Thus, the inference of "xRu" from "xRu and yRu" mentioned above corresponds to the passage from the L-formula $(xRu \wedge yRu)$ to xRu. We can view this as an application of the following *rule*:

(+) It is permissible to pass from an L-formula of the form $(\varphi \wedge \psi)$ to the L-formula φ.

In Chapter IV we shall give a finite system **S** of rules which, like (+), correspond to elementary inference steps the mathematician uses in his proofs. A *formal proof* of the L-formula ψ from the L-formulas in Φ (the "axioms") consists then (by definition) of a sequence of formulas in L which ends with ψ, and in which each L-formula is obtained by application of a rule from **S** to the axioms or to preceding formulas in the sequence.

Having introduced the precise notions, one can convince oneself by examples that mathematical proofs can be imitated by formal proofs in L. Moreover, in Chapter V we shall return to the question (*) and answer it positively, showing that if a formula ψ follows from a set Φ of formulas, then there is a proof of ψ from Φ, even a formal proof. This is the content of the so-called *Gödel completeness theorem*.

§4. Preview

Gödel's completeness theorem forms a bridge between the notion of proof, which is formal in character, and the notion of consequence, which refers to the meaning in structures. In Chapter VI we shall show how this connection can be used in algebraic investigations.

Once a formal language and an exact notion of proof have been introduced, we have a precise framework for mathematical investigations concerning, for instance, the consistency of mathematics or a justification of rules of inference used in mathematics (Chapters VII and X).

Finally, the formalization of the notion of proof creates the possibility of using a computer to carry out or check proofs. In Chapter X we shall discuss the range and the limitations of such machine-oriented methods.

In the formulas of L the variables refer to the *elements* of a structure, for example, to the elements of a group or the elements of an equivalence structure. In a given structure we often call elements of the domain A *first-order objects*, while subsets of A are called *second-order objects*. Since L only has variables for first-order objects (and thus expressions such as "$\forall x$" and "$\exists x$" apply only to the elements of a structure), we call L a *first-order language*.

Unlike L, the so-called *second-order language* also has variables which range over subsets of the domain of a structure. Thus a proposition about a given group which begins "For all subgroups..." can be directly formulated in the second-order language. We shall investigate this language and others in Chapter IX. In Chapter XII we shall be able to show that no language with more expressive power than L enjoys both an adequate formal concept of proof and other useful properties of L. From this point of view L is a "best-possible" language, and so we succeed in justifying the dominant rôle which the first-order language plays in mathematical logic.

CHAPTER II

Syntax of First-Order Languages

In this chapter we introduce the first-order languages. They obey simple, clear formation rules. In later chapters we shall discuss whether and to what extent all mathematical propositions can be formalized in such languages.

§1. Alphabets

By an *alphabet* $\mathbb{A}$ we mean a nonempty set of symbols. Examples of alphabets are the sets $\mathbb{A}_1 = \{0, 1, 2, \ldots, 9\}$, $\mathbb{A}_2 = \{a, b, c, \ldots, x, y, z\}$ (the alphabet of lower-case letters), $\mathbb{A}_3 = \{\circ, \int, a, d, x, f,), (\}$, and $\mathbb{A}_4 = \{c_0, c_1, c_2, \ldots\}$.

We call finite sequences of symbols from an alphabet $\mathbb{A}$ *strings* or *words* over $\mathbb{A}$. $\mathbb{A}^*$ denotes the set of all strings over $\mathbb{A}$. The *length* of a string $\zeta \in \mathbb{A}^*$ is the number of symbols, counting repetitions, occurring in ζ. The empty string is also considered to be a word over $\mathbb{A}$. It is denoted by $\square$, and its length is zero.

Examples of strings over $\mathbb{A}_2$ are

$$softly, \qquad xdbxaz.$$

Examples of strings over $\mathbb{A}_3$ are

$$\int f(x)\,dx, \qquad x \circ \int\int a \circ.$$

Suppose $\mathbb{A} = \{|, ||\}$, that is, $\mathbb{A}$ consists of the symbols[1] $a_1 := |$ and $a_2 := ||$. Then the string $|||$ over $\mathbb{A}$ can be read three ways: as $a_1 a_1 a_1$, as $a_1 a_2$, and as

[1] Here we write "$a_1 := |$" instead of "$a_1 = |$" in order to make it clear that a_1 is *defined* by the right-hand side of the equation.

$a_2 a_1$. In the sequel we shall allow only those alphabets $\mathbb{A}$ where any string over $\mathbb{A}$ can be read in exactly one way. The alphabets $\mathbb{A}_1, \ldots, \mathbb{A}_4$ given above satisfy this condition.

We now turn to questions concerning the number of strings over a given alphabet.

We call a set M *countable* if it is not finite and if there is a surjective map α of the set of natural numbers $\mathbb{N} = \{0, 1, 2, \ldots\}$ onto M. We can then represent M as $\{\alpha(n) | n \in \mathbb{N}\}$ or, if we write the arguments as indices, as $\{\alpha_n | n \in \mathbb{N}\}$. A set M is called *at most countable* if it is finite or countable.

1.1 Lemma. *For a nonempty set M the following are equivalent*:

(a) *M is at most countable.*
(b) *There is a surjective map* $\alpha\colon \mathbb{N} \to M$.
(c) *There is an injective map* $\beta\colon M \to \mathbb{N}$.

PROOF.[2] We shall prove (b) from (a), (c) from (b), and (a) from (c).

(b) *from* (a): Let M be at most countable. If M is countable (b) holds by definition. For finite M, say $M = \{a_0, \ldots, a_n\}$ (M is nonempty), we define $\alpha\colon \mathbb{N} \to M$ by

$$\alpha(i) := \begin{cases} a_i & \text{if } 0 \leq i \leq n, \\ a_0 & \text{otherwise.} \end{cases}$$

α is clearly surjective.

(c) *from* (b): Let $\alpha\colon \mathbb{N} \to M$ be surjective. We define an injective map $\beta\colon M \to \mathbb{N}$ by setting, for $a \in M$,

$$\beta(a) := \text{the least } i \text{ such that } \alpha(i) = a.$$

(a) *from* (c): Let $\beta\colon M \to \mathbb{N}$ be injective and suppose M is not finite. We must show that M is countable. To do this we define a surjective map $\alpha\colon \mathbb{N} \to M$ inductively as follows:

$$\begin{aligned} \alpha(0) &:= \text{the } a \in M \text{ with the smallest image under } \beta \text{ in } \mathbb{N}, \\ \alpha(n+1) &:= \text{the } a \in M \text{ with the smallest image under } \beta \text{ greater} \\ &\qquad \text{than } \beta(\alpha(0)), \ldots, \beta(\alpha(n)). \end{aligned}$$

Since the images under β are not bounded in $\mathbb{N}$, α is defined for all $n \in \mathbb{N}$, and clearly every $a \in M$ belongs to the image of α. □

Every subset of an at most countable set is at most countable. If M_1 and M_2 are at most countable, then so is $M_1 \cup M_2$. The set $\mathbb{R}$ of real numbers is neither finite nor countable: it is *uncountable*.

[2] The goal of our investigations is, among other things, a discussion of the notion of proof. Therefore the reader may be surprised that we use proofs before we have made precise what a mathematical proof is. As already mentioned in Chapter I, we shall return to this apparent circularity in Chapter VII.

We shall later show that finite alphabets suffice for representing mathematical statements. Moreover, the symbols may be chosen as "concrete" objects such that they can be included on the keyboard of a typewriter. Often, however, one can improve the transparency of an argument by using a countable alphabet such as $\mathbb{A}_4$, and we shall do this frequently. For some mathematical applications of methods of mathematical logic it is also useful to consider uncountable alphabets. The set $\{c_r \mid r \in \mathbb{R}\}$, which contains a symbol c_r for every real number r, is an example of an uncountable alphabet. We shall justify the use of such alphabets in VII.4.

1.2 Lemma. *If $\mathbb{A}$ is an at most countable alphabet, then the set $\mathbb{A}^*$ of strings over $\mathbb{A}$ is countable.*

PROOF. Let p_n be the nth prime number: $p_0 = 2$, $p_1 = 3$, $p_2 = 5$, and so on. If $\mathbb{A}$ is finite, say $\mathbb{A} = \{a_0, \ldots, a_n\}$, where $a_0, \ldots, a_n$ are pairwise distinct, or if $\mathbb{A}$ is countable, say $\mathbb{A} = \{a_0, a_1, a_2, \ldots\}$, where the a_i are pairwise distinct, we can define the map $\beta: \mathbb{A}^* \to \mathbb{N}$ by

$$\beta(\square) = 1, \qquad \beta(a_{i_0} \ldots a_{i_r}) = p_0^{i_0+1} \cdot \ldots \cdot p_r^{i_r+1}.$$

Clearly β is injective and thus (cf. 1.1(c)) $\mathbb{A}^*$ is at most countable. Since $a_0, a_0 a_0, a_0 a_0 a_0, \ldots$ are all in $\mathbb{A}^*$ it cannot be finite; hence it is countable. $\square$

1.3 Exercise. Let $\alpha: \mathbb{N} \to \mathbb{R}$ be given. For $a, b \in \mathbb{R}$ such that $a < b$ show that there is a point c in the interval $I = [a, b]$ such that $c \notin \{\alpha(n) \mid n \in \mathbb{N}\}$. Conclude from this that I, and hence $\mathbb{R}$ also, are uncountable. (*Hint*: By induction define a sequence $I = I_0 \supset I_1 \supset \ldots$ of closed intervals such that $\alpha(n) \notin I_{n+1}$, and use the fact that $\bigcap_{n \in \mathbb{N}} I_n \neq \emptyset$.)

1.4 Exercise. Show that if M_1 and M_3 are countable sets and $M_1 \subset M_2 \subset M_3$, then M_2 is also countable.

1.5 Exercise. (a) Show that if the sets $M_0, M_1, \ldots$ are at most countable then the union $\bigcup_{i \in \mathbb{N}} M_i$ is also at most countable.

(b) Use (a) to give a different proof of Lemma 1.2.

§2. The Alphabet of a First-Order Language

We wish to construct formal languages in which we can formulate, for example, the axioms, theorems, and proofs about groups and equivalence relations which we considered in Chapter I. In that context the connectives, the quantifiers, and the equality relation played an important rôle. Therefore, we shall include the following symbols in the first-order languages: $\neg$ (for "not"), $\wedge$ (for "and"), $\vee$ (for "or"), $\rightarrow$ (for "if-then"), $\leftrightarrow$ (for "if and only if"), $\forall$ (for "for all"), $\exists$ (for "there exists"), $\equiv$ (as symbol for equality). To these we shall add variables (for elements of groups, elements of equivalence structures, etc.) and finally parentheses as auxiliary symbols. In order to formulate the axioms for groups we also need certain symbols peculiar to group theory, e.g., a *binary function symbol*, say $\circ$, to denote the group multiplication, and a symbol, say e, to denote the identity element. We call e a constant symbol, or simply, a *constant*. For the axioms of the theory of equivalence relations we need a *binary relation symbol*, say R.

Thus, in addition to the "logical" symbols such as "$\neg$" and "$\wedge$", we shall need a set S of relation symbols, function symbols, and constants which varies from theory to theory. Each such set S of symbols determines a first-order language.

We summarize:

2.1 Definition. The *alphabet of a first-order language* contains the following symbols:

(a) $v_0, v_1, v_2, \ldots$ (*variables*);
(b) $\neg, \wedge, \vee, \rightarrow, \leftrightarrow$ (*not, and, or, if-then, if and only if*);
(c) $\forall, \exists$ (*for all, there exists*);
(d) $\equiv$ (*equality symbol*);
(e)), ((*parentheses*);
(f) (1) for every $n \geq 1$ a (possibly empty) set of *n-ary relation symbols*;
 (2) for every $n \geq 1$ a (possibly empty) set of *n-ary function symbols*;
 (3) a (possibly empty) set of *constants*.

We shall denote by $\mathbb{A}$ the set of symbols listed in (a) through (e), and by S the set of symbols from (f). S may be empty. The symbols listed under (f) must, of course, be distinct from each other and from the symbols in $\mathbb{A}$.

S determines a first-order language (cf. §3). We call $\mathbb{A}_S := \mathbb{A} \cup S$ the alphabet of this language and S its *symbol set*.

We have already become acquainted with some symbol sets: $S_{\text{gr}} := \{\circ, e\}$ for group theory and $S_{\text{eq}} := \{R\}$ for the theory of equivalence relations. For the theory of ordered groups we could use $\{\circ, e, R\}$, where the binary relation symbol R is now taken to represent the ordering relation. In certain theoretical investigations we shall use the symbol set S_∞, which contains the constants $c_0, c_1, c_2, \ldots$, and for every $n \geq 1$ the countably many n-ary relation symbols $R_0^n, R_1^n, R_2^n, \ldots$ and n-ary function symbols $f_0^n, f_1^n, f_2^n, \ldots$.

Henceforth we shall use the letters $P, Q, R, \ldots$ to stand for relation symbols, $f, g, h, \ldots$ for function symbols, $c, c_0, c_1, \ldots$ for constants, and $x, y, z, \ldots$ for variables.

§3. Terms and Formulas in First-Order Languages

Given a symbol set S, we call certain strings over $\mathbb{A}_S$ *formulas* of the first-order language determined by S. For example, if $S = S_{\mathrm{gr}}$ we want the strings

$$e \equiv e, \qquad e \circ v_1 \equiv v_2, \qquad \exists v_1(e \equiv e \wedge v_1 \equiv v_2)$$

to be formulas, but not

$$\equiv \wedge\, e, \qquad e \vee e.$$

The formulas $e \equiv e$ and $e \circ v_1 \equiv v_2$ have the form of equations. Mathematicians call the strings to the left and to the right of the equality symbol *terms*. Terms are "meaningful" combinations of function symbols, variables, and constants (together with commas and parentheses). Clearly, to give a precise definition of formulas and thus, in particular, of equations, we must first specify more exactly what we mean by terms. In mathematics terms are written in different notations, such as $x + e$, $g(x, e)$, gxe. We choose a parenthesis-free notation system, as in gxe.

To define the notion of term we give *instructions* (or *rules*) which tell us how to generate the terms. (Such a system of rules is often called a *calculus*.) This is more precise than a vague description, and simpler than an explicit definition.

3.1 Definition. *S-terms* are precisely those strings in $\mathbb{A}_S^*$ which can be obtained by finitely many applications of the following rules:

(T1) Every variable is an S-term.
(T2) Every constant in S is an S-term.
(T3) If the strings $t_0, \ldots, t_{n-1}$ are S-terms and f is an n-ary function symbol in S, then $ft_0 \ldots t_{n-1}$ is also an S-term.

We denote the set of S-terms by T^S.

If f is a unary and g a binary function symbol and $S = \{f, g, c, R\}$, then

$$gv_0\, fgv_4c$$

is an S-term. First of all, c is an S-term by (T2) and v_0 and v_4 are S-terms by (T1). If we apply (T3) to the S-terms v_4 and c and to the function symbol g, we see that gv_4c is an S-term. Another application of (T3) to the S-term gv_4c and to the function symbol f shows that fgv_4c is an S-term, and a final application of (T3) to the S-terms v_0 and fgv_4c and to the function symbol g shows that $gv_0\, fgv_4c$ is an S-term.

We say that one can *derive* the string $gv_0\, fgv_4c$ in the calculus of terms (corresponding to S). The *derivation* just described can be given schematically as follows:

1. c (T2)
2. v_0 (T1)
3. v_4 (T1)
4. gv_4c (T3) applied to 3 and 1 using g
5. fgv_4c (T3) applied to 4 using f
6. gv_0fgv_4c (T3) applied to 2 and 5 using g.

The string directly following the number at the beginning of each line can be obtained in each case by applying a rule of the calculus of terms; applications of (T3) use terms obtained in preceding lines. The information at the end of each line indicates which rules and preceding terms were used. Clearly, not only the string in the last line, but all strings in preceding lines can be derived and hence are S-terms.

The reader should show that the strings $gxgxfy$ and $gxgfxfy$ are S-terms for arbitrary variables x and y. Here we give a derivation to show that the string $\circ x \circ ey$ is an S_{gr}-term.

1. x (T1)
2. y (T1)
3. e (T2)
4. $\circ ey$ (T3) applied to 3 and 2 using $\circ$
5. $\circ x \circ ey$ (T3) applied to 1 and 4 using $\circ$.

Mathematicians usually write the term in line 4 as $e \circ y$, and the term in line 5 as $x \circ (e \circ y)$. For easier reading we shall sometimes write terms in this way as well.

Using the notion of term we are now able to give the definition of formulas.

3.2 Definition. *S-formulas* are precisely those strings of $\mathbb{A}_S^*$ which are obtained by finitely many applications of the following rules:

(F1) If t_0 and t_1 are S-terms, then $t_0 \equiv t_1$ is an S-formula.

(F2) If $t_0, \ldots, t_{n-1}$ are S-terms and R is an n-ary relation symbol from S, then $Rt_0 \ldots t_{n-1}$ is an S-formula.

(F3) If φ is an S-formula, then $\neg\varphi$ is also an S-formula.

(F4) If φ and ψ are S-formulas, then $(\varphi \wedge \psi)$, $(\varphi \vee \psi)$, $(\varphi \rightarrow \psi)$, and $(\varphi \leftrightarrow \psi)$ are also S-formulas.

(F5) If φ is an S-formula and x is a variable, then $\forall x\varphi$ and $\exists x\varphi$ are also S-formulas.

S-formulas derived using (F1) and (F2) are called *atomic formulas* because they are not formed by combining other S-formulas. $\neg\varphi$ is called the *negation* of φ, and $(\varphi \wedge \psi)$, $(\varphi \vee \psi)$, and $(\varphi \rightarrow \psi)$ are called, respectively, the *conjunction*, *disjunction*, and *implication* of φ and ψ.

We use L^S to denote the set of S-formulas. L^S is the *first-order language corresponding to the symbol set S* (often called the *language of first-order predicate calculus* corresponding to S).

Instead of S-terms and S-formulas, we often speak simply of terms and formulas when the reference to S is either clear or unimportant. For terms we use the letters $t, t_0, t_1, \ldots$, and for formulas the letters $\varphi, \psi, \ldots$.

We now give some examples. Let $S = S_{eq} = \{R\}$. We can express the axioms for the theory of equivalence relations by the following formulas:

$$\forall v_0 R v_0 v_0,$$

$$\forall v_0 \forall v_1 (R v_0 v_1 \to R v_1 v_0),$$

$$\forall v_0 \forall v_1 \forall v_2 ((R v_0 v_1 \wedge R v_1 v_2) \to R v_0 v_2).$$

One can verify that these strings really are formulas by giving appropriate derivations (as was done above for terms) in the calculus of S_{eq}-formulas. For the first two formulas we have, for example,

(1) 1. $R v_0 v_0$ (F2)
    2. $\forall v_0 R v_0 v_0$ (F5) applied to 1 using $\forall$, v_0.

(2) 1. $R v_0 v_1$ (F2)
    2. $R v_1 v_0$ (F2)
    3. $(R v_0 v_1 \to R v_1 v_0)$ (F4) applied to 1 and 2 using $\to$
    4. $\forall v_1 (R v_0 v_1 \to R v_1 v_0)$ (F5) applied to 3 using $\forall$, v_1
    5. $\forall v_0 \forall v_1 (R v_0 v_1 \to R v_1 v_0)$ (F5) applied to 4 using $\forall$, v_0.

In a similar way the reader should convince himself that, for unary f, binary g, unary P, ternary Q, and variables x, y, and z, the following strings are $\{f, g, P, Q\}$-formulas:

(1) $\forall y (Pz \to Qxxz)$;
(2) $(Pgxfy \to \exists x(x \equiv x \wedge x \equiv x))$;
(3) $\forall z \forall z \exists z\, Qxyz$.

In spite of its rigor the calculus of formulas has "liberal" aspects: we can quantify over a variable which does not actually occur in the formula in question (as in (1)), we can join two identical formulas by means of a conjunction (as in (2)), or we can quantify several times over the same variable (as in (3)).

For the sake of clarity we shall frequently use an abbreviated or modified notation for terms and formulas. For example, we shall write the S_{eq}-formula $R v_0 v_1$ as $v_0 R v_1$ (compare this with the notation $2 < 3$). Moreover, we shall often omit parentheses if they are not essential in order to avoid ambiguity, e.g., the outermost parentheses surrounding conjunctions, disjunctions, etc. Thus we may write $\varphi \wedge \psi$ for $(\varphi \wedge \psi)$. In the case of iterated conjunctions or disjunctions we shall agree to associate to the left, e.g., $\varphi \wedge \psi \wedge \chi$ will be understood to mean $((\varphi \wedge \psi) \wedge \chi)$. The reader should always be aware that expressions in the abbreviated form are no longer formulas in the sense of 3.2. We emphasize once again that we need an exact definition of formulas in order to have a precise notion of mathematical statement in our analysis of the notion of proof.

Perhaps the following analogy with programming languages will clarify the situation. When writing a program one must be meticulous in following the grammatical rules for the programming language, because a computer can process only a formally correct program. But programmers use an abbreviated notation when devising or discussing programs in order to express themselves more quickly and clearly.

We have used $\equiv$ for the equality symbol in first-order languages in order to make statements of the form $\varphi = x \equiv y$ ("φ is the formula $x \equiv y$") easier to read.

For future use we note the following:

3.3 Lemma. *If S is at most countable, then T^S and L^S are countable.*

PROOF. If S is at most countable, then so is $\mathbb{A}_S$ and hence by 1.2 the set $\mathbb{A}_S^*$. Since T^S and L^S are subsets of $\mathbb{A}_S^*$ they are also at most countable. On the other hand T^S and L^S are infinite because T^S contains the variables $v_0, v_1, v_2, \ldots$, and L^S contains the formulas $v_0 \equiv v_0, v_1 \equiv v_1, v_2 \equiv v_2, \ldots$ (even if $S = \emptyset$). □

With the preceding observations the languages L^S have become the object of investigations. In these investigations we use another language, namely everyday English augmented by some mathematical terminology. In order to emphasize the difference in the present context, the formal language L^S, being discussed, is called the *object language*; the language English, in which we discuss, is called the *metalanguage*. In another context, for example, in linguistic investigations, everyday English could be an object language. Similarly, first-order languages can play the rôle of metalanguages in certain set-theoretical investigations (cf. VII.4.3).

Historical Note. G. Frege [10] developed the first comprehensive formal language. He used a two-dimensional system of notation which was so complicated that his language never came into general use. The formal languages used today are based essentially on those introduced by G. Peano [22].

§4. Induction in the Calculus of Terms and in the Calculus of Formulas

Let S be a set of symbols and let $Z \subset \mathbb{A}_S^*$ be a set of strings over $\mathbb{A}_S$. In the case where $Z = T^S$ or $Z = L^S$ we described the elements of Z by means of a calculus. Each rule of such a calculus either says that certain strings belong to Z (e.g., the rules (T1), (T2), (F1), and (F2)), or else permits the passage from certain strings $\zeta_0, \ldots, \zeta_{n-1}$ to a new string ζ in the sense that, if $\zeta_0, \ldots, \zeta_{n-1}$ all belong to Z, then ζ also belongs to Z. The way such rules work is made clear when we write them schematically, as follows:

$$\frac{\zeta_0, \ldots, \zeta_{n-1}}{\zeta}.$$

By allowing $n = 0$, the first sort of rule mentioned above ("premise-free" rules) is included in this scheme.

Now we can write the rules for the calculus of terms as follows:

(T1) $\dfrac{}{x}$;

(T2) $\dfrac{}{c}$, if $c \in S$;

(T3) $\dfrac{t_0, \ldots, t_{n-1}}{ft_0 \ldots t_{n-1}}$, if $f \in S$ and f is n-ary.

When we define a set Z of strings by means of a calculus **C** we can then prove assertions about the elements of Z by means of *induction over* **C**. This principle of proof corresponds to induction over the natural numbers. If one wants to show that all elements of Z have a certain property P, then it is sufficient to show that

(I) for every rule

(*) $$\frac{\zeta_0, \ldots, \zeta_{n-1}}{\zeta}$$

of the calculus **C**, the following holds: whenever $\zeta_0, \ldots, \zeta_{n-1}$ are derivable in **C** and have the property P ("*induction hypothesis*"), then ζ also has the property P.

Hence in case $n = 0$ we must show that ζ has the property P.

This principle of proof is evident: in order to show that all strings derivable in **C** have the property P, we show that everything derivable by means of a "premise-free" (i.e., $n = 0$ in (*)) rule has the property P, and then that P is preserved under application of the remaining rules. The method can also be justified using the principle of complete induction for natural numbers. For this purpose, one defines in the obvious way the length of a derivation in **C** (cf. the examples of derivations in §3), and then argues as follows: If the condition (I) is satisfied for P, one shows by induction on n that every string which has a derivation of length n has the property P. Since every element of Z has a derivation of some finite length, P must then hold for all elements of Z.

In the special case where **C** is the calculus of terms or the calculus of formulas, we call the proof procedure outlined above *proof by induction on* *terms* or *on formulas*, respectively. In order to show that all S-terms have a certain property P it is sufficient to show:

(T1)′ Every variable has the property P.
(T2)′ Every constant in S has the property P.
(T3)′ If the S-terms $t_0, \ldots, t_{n-1}$ have the property P, and if $f \in S$ is n-ary, then $ft_0 \ldots t_{n-1}$ also has the property P.

In the case of the calculus of formulas the corresponding conditions are

(F1)′ Every S-formula of the form $t_0 \equiv t_1$ has the property P.
(F2)′ Every S-formula of the form $Rt_0 \ldots t_{n-1}$ has the property P.
(F3)′ If the S-formula φ has the property P, then $\neg\varphi$ also has the property P.
(F4)′ If the S-formulas φ and ψ have the property P, then $(\varphi \wedge \psi)$, $(\varphi \vee \psi)$, $(\varphi \rightarrow \psi)$, and $(\varphi \leftrightarrow \psi)$ also have the property P.
(F5)′ If the S-formula φ has the property P and if x is a variable, then $\forall x\varphi$ and $\exists x\varphi$ also have the property P.

We now give some applications of this method of proof.

4.1 Lemma. (a) *For all symbol sets S the empty string* $\square$ *is neither an S-term nor an S-formula.*
(b) (1) $\circ$ *is not an* S_{gr}*-term.*
(2) $\circ\circ v_1$ *is not an* S_{gr}*-term.*
(c) *For all symbol sets S, every S-formula contains the same number of right parentheses* "$)$" *as left parentheses* "$($".

PROOF. (a) Let P be the property on $\mathbb{A}_S^*$ which holds for a string ζ iff ζ is nonempty. We show by induction on terms that every S-term has the property P, and leave the proof for formulas to the reader.

(T1)′ and (T2)′: Terms of the form x or c are nonempty.

(T3)′: Every term formed according to (T3) begins with a function symbol, and hence is nonempty. (Note that we do not need to use the induction hypothesis.)

(b) We leave (1) to the reader. To prove (2), let P be the property on $\mathbb{A}_{S_{\mathrm{gr}}}^*$ which holds for a string ζ over $\mathbb{A}_{S_{\mathrm{gr}}}$ iff ζ is distinct from $\circ\circ v_1$. We shall show by induction on terms that every S_{gr}-term is distinct from $\circ\circ v_1$. The reader will notice that we start using a more informal presentation of inductive proofs.

$t = x$, $t = e$: t is distinct from the string $\circ\circ v_1$.

$t = \circ t_1 t_2$: If $\circ t_1 t_2$ were $\circ\circ v_1$ then, by (a), we would have $t_1 = \circ$ and $t_2 = v_1$. But $t_1 = \circ$ contradicts (1).

(c) First one shows by induction on terms that no S-term contains a left or right parenthesis. Then one considers the property P over $\mathbb{A}_S^*$, which holds for a string ζ over $\mathbb{A}_S$ iff ζ has the same number of left parentheses as right parentheses, and one shows by induction on formulas that every S-formula has the property P. We give some cases here as examples:

$\varphi = t_0 \equiv t_1$, where t_0 and t_1 are S-terms: By the observation above there are no parentheses in φ, thus P holds for φ.

$\varphi = \neg\psi$, where ψ has the property P by induction hypothesis: Since φ does not contain any parentheses except those in ψ, φ also has the property P.

$\varphi = (\psi \wedge \chi)$, where P holds for ψ and χ by induction hypothesis: Since φ contains one left parenthesis and one right parenthesis in addition to the parentheses in ψ and χ, the property P also holds for φ.

$\varphi = \forall x\psi$, where ψ has the property P by induction hypothesis: The proof here is the same as in the case $\varphi = \neg\psi$. □

Next we want to show that terms and formulas have a unique decomposition into their constituents. The following two lemmas contain some preliminary results needed for this purpose. We refer to a fixed symbol set S.

4.2 Lemma. (a) *For all terms t and t', t is not a proper initial segment of t' (i.e., there is no ζ distinct from $\square$ such that $t\zeta = t'$).*
(b) *For all formulas φ and φ', φ is not a proper initial segment of φ'.*

We confine ourselves to the *proof of* (a), and consider the property P, which holds for a string η iff

(∗) for all terms t', η is not a proper initial segment of t' and t' is not a proper initial segment of η.

Using induction on terms, we show that all terms t have the property P. (a) will then be proved.

$t = x$: Let t' be an arbitrary term. By 4.1(a), t' cannot be a proper initial segment of x, for then t' would have to be the empty string $\square$. On the other hand, one can easily show by induction on terms that x is the only term which begins with the variable x. Therefore, t cannot be a proper initial segment of t'.

$t = c$: The argument is similar.

$t = ft_0 \dots t_{n-1}$ and (∗) holds for $t_0, \dots, t_{n-1}$: Let t' be an arbitrary fixed term. We show that t' cannot be a proper initial segment of t. Otherwise there would be a ζ such that

(1) $$\zeta \neq \square, \qquad t = t'\zeta.$$

Since t' begins with f (for t begins with f), t' cannot be a variable or a constant, thus t' must have been generated using (T3). Therefore it has the form $ft'_0 \dots t'_{n-1}$ for suitable terms $t'_0 \dots t'_{n-1}$. From (1) we have

(2) $$ft_0 \dots t_{n-1} = ft'_0 \dots t'_{n-1}\zeta,$$

and from this, cancelling the symbol f, we obtain

(3) $$t_0 \dots t_{n-1} = t'_0 \dots t'_{n-1}\zeta.$$

Therefore t_0 is an initial segment of t'_0 or vice versa. Since t_0 satisfies (*) by induction hypothesis, neither of these can be a proper initial segment of the other. Thus $t_0 = t'_0$. Cancelling t_0 on both sides of (3) we obtain

(4) $$t_1 \dots t_{n-1} = t'_1 \dots t'_{n-1}\zeta.$$

Repeatedly applying the argument leading from (3) to (4) we finally obtain

$$\square = \zeta.$$

This contradicts (1). Therefore t' cannot be a proper initial segment of t. The proof that t cannot be a proper initial segment of t' is analogous. □

Applying 4.2 one can obtain in a similar way

4.3 Lemma. (a) *If $t_0, \dots, t_{n-1}$ and $t'_0, \dots, t'_{m-1}$ are terms, and if $t_0 \dots t_{n-1} = t'_0 \dots t'_{m-1}$, then $n = m$ and $t_i = t'_i$ for $i < n$.*
(b) *If $\varphi_0, \dots, \varphi_{n-1}$ and $\varphi'_0, \dots, \varphi'_{m-1}$ are formulas, and if $\varphi_0 \dots \varphi_{n-1} = \varphi'_0 \dots \varphi'_{m-1}$, then $n = m$ and $\varphi_i = \varphi'_i$ for $i < n$.*

Using 4.2 and 4.3 one can then easily prove

4.4 Theorem. (a) *Every term is either a variable or a constant or a term of the form $ft_0 \dots t_{n-1}$. In the last case the function symbol f and the terms $t_0, \dots, t_{n-1}$ are uniquely determined.*
(b) *Every formula is of the form* (1) $t_0 \equiv t_1$, *or* (2) $Rt_0 \dots t_{n-1}$, *or* (3) $\neg\varphi$, *or* (4) $(\varphi \wedge \psi)$, *or* (5) $(\varphi \vee \psi)$, *or* (6) $(\varphi \rightarrow \psi)$, *or* (7) $(\varphi \leftrightarrow \psi)$, *or* (8) $\forall x\varphi$, *or* (9) $\exists x\varphi$, *where the cases* (1)–(9) *are mutually exclusive and the following are uniquely determined: the terms t_0 and t_1 in case* (1), *the relation symbol R and the terms $t_0, \dots, t_{n-1}$ in case* (2), *the formula φ in* (3), *the formulas φ and ψ in* (4), (5), (6), *and* (7), *and the variable x and the formula φ in* (8) *and* (9). □

Theorem 4.4 asserts that a term or a formula has a unique decomposition into its constituents. Thus, as we shall now show, we can give *inductive definitions on terms* or *formulas*. For example, to define a function for all terms it will be sufficient

(T1)″ to assign a value to each variable;
(T2)″ to assign a value to each constant;
(T3)″ for every n-ary f and for all terms $t_0, \dots, t_{n-1}$ to assign a value to the term $ft_0 \dots t_{n-1}$ assuming that values have already been assigned to the terms $t_0, \dots, t_{n-1}$.

Each term is assigned exactly one value by (T1)″ through (T3)″. We show this by means of induction on terms as follows.

$t = x$: By 4.4(a) t is not a constant and does not begin with a function symbol. Therefore, it is assigned a value only by an application of (T1)″. Thus t is assigned exactly one value.

$t = c$: The argument is analogous to the preceding case.

$t = ft_0 \ldots t_{n-1}$, *and each of the terms* $t_0, \ldots, t_{n-1}$ *has been assigned exactly one value*: To assign a value to t we can only use (T3)″, by 4.4(a). Since, again by 4.4(a), the t_i are uniquely determined, t is assigned a unique value. □

We now give some examples of inductive definitions.

The function var (more precisely, var_S), which associates with each S-term the set of variables occurring in it, can be defined as follows:

4.5 Definition.

$$\mathrm{var}(x) := \{x\};$$

$$\mathrm{var}(c) := \varnothing;$$

$$\mathrm{var}(ft_0 \ldots t_{n-1}) := \mathrm{var}(t_0) \cup \cdots \cup \mathrm{var}(t_{n-1}).$$

The function SF, which assigns to each formula the set of its subformulas, can be defined by induction on formulas as follows:

$$\mathrm{SF}(t_0 \equiv t_1) := \{t_0 \equiv t_1\};$$

$$\mathrm{SF}(Rt_0 \ldots t_{n-1}) := \{Rt_0 \ldots t_{n-1}\};$$

$$\mathrm{SF}(\neg\varphi) := \{\neg\varphi\} \cup \mathrm{SF}(\varphi);$$

$$\mathrm{SF}((\varphi \wedge \psi)) := \{(\varphi \wedge \psi)\} \cup \mathrm{SF}(\varphi) \cup \mathrm{SF}(\psi);$$

similarly for $(\varphi \vee \psi)$, $(\varphi \rightarrow \psi)$, and $(\varphi \leftrightarrow \psi)$;

$$\mathrm{SF}(\forall x\varphi) := \{\forall x\varphi\} \cup \mathrm{SF}(\varphi);$$

similarly for $\exists x\varphi$.

A means of defining the preceding notions by calculi is indicated in the following exercise.

4.6 Exercise. (a) Let the calculus $\mathbf{C}_v$ consist of the following rules:

$$\frac{}{x \quad x}; \qquad \frac{y \quad t_i}{y \quad ft_0 \ldots t_{n-1}}, \quad \text{if } f \in S \text{ is } n\text{-ary and } i < n.$$

Show that, for all variables x and all S-terms t, xt is derivable in $\mathbf{C}_v$ iff $x \in \mathrm{var}(t)$.

(b) Give a result for SF analogous to the result for var in (a).

4.7 Exercise. Alter the calculus of formulas by omitting the parentheses in 3.2(F4), e.g., by writing simply "$\varphi \wedge \psi$" instead of "$(\varphi \wedge \psi)$". So, for example, $\chi := \exists v_0 Pv_0 \wedge Qv_1$ is a $\{P, Q\}$-formula in this new sense. Show that the analogue of 4.4 no longer holds, and that the corresponding definition of

SF(χ) can yield both SF(χ) $= \{\chi, Pv_0 \wedge Qv_1, Pv_0, Qv_1\}$ and SF(χ) $=$ $(\chi, \exists v_0\, Pv_0, Pv_0, Qv_1\}$, so that SF is no longer a well-defined function.

4.8 Exercise (Parenthesis-Free, or So-Called Polish Notation for Formulas). Let S be a symbol set and let $\mathbb{A}'$ be the set of symbols given in 2.1(a)–(d). Let $\mathbb{A}'_S = \mathbb{A}' \cup S$. Define S-formulas in Polish notation ($S - P$-formulas) to be all strings over $\mathbb{A}'_S$ which can be obtained by finitely many applications of the rules (F1), (F2), (F3), and (F5) from 3.2, and the rule (F4)$'$:

(F4)$'$ If φ an ψ are $S - P$-formulas then $\wedge\varphi\psi$, $\vee\varphi\psi$, $\rightarrow\varphi\psi$, and $\leftrightarrow\varphi\psi$ are also $S - P$-formulas.

Prove the analogues of 4.3(b) and 4.4(b) for $S - P$-formulas.

§5. Free Variables and Sentences

Let x, y, and z be distinct variables. In the $\{R\}$-formula

$$\varphi := \exists x(R\underline{y}\underline{z} \wedge \forall y(\neg \underline{\underline{y}} \equiv \underline{\underline{x}} \vee R\underline{\underline{y}}\underline{z}))$$

the occurrences of the variables y and z marked with single underlining are not quantified, i.e., not in the scope of a corresponding quantifier. Such occurrences are called *free*, and as we shall see later, the variables there act as *parameters*. The occurrences of the variables x and y marked with double underlining shall be called *bound* occurrences. (Thus y has both free and bound occurrences in φ.)

We give a definition by induction on formulas of the set of *free variables in a formula* φ; we denote this set by free(φ). Again we fix a symbol set S.

5.1 Definition.

$$\text{free}(t_0 \equiv t_1) := \text{var}(t_0) \cup \text{var}(t_1);$$

$$\text{free}(Pt_0 \ldots t_{n-1}) := \text{var}(t_0) \cup \cdots \cup \text{var}(t_{n-1});$$

$$\text{free}(\neg\varphi) := \text{free}(\varphi);$$

$$\text{free}((\varphi * \psi)) := \text{free}(\varphi) \cup \text{free}(\psi) \quad \text{for } * = \wedge, \vee, \rightarrow, \leftrightarrow;$$

$$\text{free}(\forall x\varphi) := \text{free}(\varphi) - \{x\};$$

$$\text{free}(\exists x\varphi) := \text{free}(\varphi) - \{x\}.$$

The reader should use this definition to determine the set of free variables in the formula φ at the beginning of this section ($S = \{R\}$). We do this here for a simpler example. Again let x, y, and z be distinct variables.

$$\begin{aligned}\text{free}((Ryx \rightarrow \forall y \neg y \equiv z)) &= \text{free}(Ryx) \cup \text{free}(\forall y \neg y \equiv z) \\ &= \{x, y\} \cup (\text{free}(\neg y \equiv z) - \{y\}) \\ &= \{x, y\} \cup (\{y, z\} - \{y\}) = \{x, y, z\}.\end{aligned}$$

Formulas without free variables ("parameter-free" formulas) are called *sentences*. For example, $\exists v_0 \neg v_0 \equiv v_0$ is a sentence.

Finally, we denote by L_n^S the set of S-formulas in which the variables occurring free are among $v_0, \ldots, v_{n-1}$:

$$L_n^S := \{\varphi \mid \varphi \text{ is an } S\text{-formula and free } (\varphi) \subset \{v_0, \ldots, v_{n-1}\}\}.$$

In particular L_0^S is the set of L-sentences.

5.2 Exercise. Show that the following calculus $\mathbf{C}_{\text{nf}}$ permits to derive precisely those strings of the form $x\varphi$ for which $\varphi \in L^S$ and for which x does not occur free in φ.

$$\frac{}{x \quad t_0 \equiv t_1}, \quad \text{if } t_0, t_1 \in T^S \text{ and } x \notin \text{var}(t_0) \cup \text{var}(t_1);$$

$$\frac{}{x \;\; Rt_0 \ldots t_{n-1}}, \quad \begin{array}{l} \text{if } R \in S \text{ is } n\text{-ary, } t_0, \ldots, t_{n-1} \in T^S, \text{ and} \\ x \notin \text{var}(t_0) \cup \cdots \cup \text{var}(t_{n-1}); \end{array}$$

$$\frac{x \quad \varphi}{x \neg \varphi}; \qquad \frac{\begin{array}{cc} x & \varphi \\ x & \psi \end{array}}{x(\varphi * \psi)} \quad \text{for } * = \wedge, \vee, \rightarrow, \leftrightarrow;$$

$$\frac{}{x \; \forall x \varphi}; \qquad \frac{}{x \; \exists x \varphi};$$

$$\frac{x \quad \varphi}{x \; \forall y \varphi}, \quad \text{if } x \neq y; \qquad \frac{x \quad \varphi}{x \; \exists y \varphi}, \quad \text{if } x \neq y.$$

CHAPTER III
Semantics of First-Order Languages

Let R be a binary relation symbol. The $\{R\}$-formula

$$(1) \qquad \forall v_0 \, R v_0 v_0$$

is, at present, merely a string of symbols to which no meaning is attached. The situation changes if we specify a domain for the variable v_0 and if we interpret the binary relation symbol R as a binary relation over this domain. There are, of course, many possible choices for such a domain and relation.

For example, suppose we choose $\mathbb{N}$ for the domain, take "$\forall v_0$" to mean "for all $n \in \mathbb{N}$" and interpret R as the divisibility relation $R^{\mathbb{N}}$ on $\mathbb{N}$. Then clearly (1) becomes the (true) statement

$$\textit{for all } n \in \mathbb{N}, \; R^{\mathbb{N}}nn,$$

i.e., the statement

every natural number is divisible by itself.

We say that the formula $\forall v_0 \, R v_0 v_0$ *holds in* $(\mathbb{N}, R^{\mathbb{N}})$. But if we choose the set $\mathbb{Z}$ of integers as the domain and interpret R as the "smaller-than" relation $R^{\mathbb{Z}}$ on $\mathbb{Z}$, then (1) becomes the (false) statement

$$\textit{for all } a \in \mathbb{Z}, \; R^{\mathbb{Z}}aa,$$

i.e., the statement

for every integer a, $a < a$.

We say that the formula $\forall v_0 \, R v_0 v_0$ *does not hold in* $(\mathbb{Z}, R^{\mathbb{Z}})$. If we consider the formula

$$\exists v_0 (R v_1 v_0 \wedge R v_0 v_2)$$

in $(\mathbb{Z}, R^{\mathbb{Z}})$, we must also interpret the free variables v_1 and v_2 as elements of $\mathbb{Z}$. If we interpret v_1 as 5 and v_2 as 8 we obtain the (true) statement

there is an integer a such that $5 < a$ *and* $a < 8$.

If we interpret v_1 as 5 and v_2 as 6 we get the (false) statement

there is an integer a such that $5 < a$ *and* $a < 6$.

The central aim of this chapter is to give a rigorous formulation of the notion of interpretation and to define precisely when an interpretation yields a true (or false) statement. We shall then be able to define in an exact way the consequence relation, which was mentioned in Chapter I.

The definitions of "term", "formula", "free occurrence", etc., given in Chapter II, involve only formal (i.e., grammatical) properties of symbol strings. We call these concepts *syntactic*. On the other hand the concepts which we shall introduce in this chapter will depend on the *meanings* of symbol strings also (for example, on the meaning in structures, as in the case above). Such concepts are called *semantic concepts*.

§1. Structures and Interpretations

If A is a set and $n \geq 1$, an *n-ary function on A* is a map whose domain of definition is the set A^n of n-tuples of elements from A, and whose values lie in A. By an *n-ary relation* $\mathfrak{Q}$ *on A* we mean a subset of A^n. Instead of writing $(a_0, \ldots, a_{n-1}) \in \mathfrak{Q}$, we shall often write $\mathfrak{Q}a_0 \ldots a_{n-1}$, and we shall say that the relation $\mathfrak{Q}$ holds for $a_0, \ldots, a_{n-1}$.

According to this definition the relation "smaller-than" on $\mathbb{Z}$ is the set $\{(a, b) \mid a, b \in \mathbb{Z} \text{ and } a < b\}$.

In the examples given earlier, the structures $(\mathbb{N}, R^{\mathbb{N}})$ and $(\mathbb{Z}, R^{\mathbb{Z}})$ were determined by the domains $\mathbb{N}$ and $\mathbb{Z}$ and by the binary relations $R^{\mathbb{N}}$ and $R^{\mathbb{Z}}$ as interpretations of the symbol R. We call $(\mathbb{N}, R^{\mathbb{N}})$ and $(\mathbb{Z}, R^{\mathbb{Z}})$ $\{R\}$-structures, thereby specifying the set of interpreted symbols, in this case $\{R\}$.

Consider once more the symbol set $S_{\text{gr}} = \{\circ, e\}$ of group theory. If we take the real numbers $\mathbb{R}$ as the domain and interpret $\circ$ as addition over $\mathbb{R}$ and e as the element 0 of $\mathbb{R}$, then we obtain the S_{gr}-structure $(\mathbb{R}, +, 0)$. In general an S-structure $\mathfrak{A}$ is determined by specifying:

(a) a domain A,
(b) (1) an n-ary relation on A for every n-ary relation symbol in S,
 (2) an n-ary function on A for every n-ary function symbol in S,
 (3) an element of A for every constant in S.

We combine the separate parts of (b) and give:

1.1 Definition. An *S-structure* is a pair $\mathfrak{A} = (A, \mathfrak{a})$ with the following properties:

(a) A is a *nonempty set*, the *domain* or *universe* of $\mathfrak{A}$.
(b) $\mathfrak{a}$ is a map defined on S satisfying:
  (1) for every n-ary relation symbol R in S, $\mathfrak{a}(R)$ is an n-ary relation on A,
  (2) for every n-ary function symbol f in S, $\mathfrak{a}(f)$ is an n-ary function on A,
  (3) for every constant c in S, $\mathfrak{a}(c)$ is an element of A.

Instead of $\mathfrak{a}(R)$, $\mathfrak{a}(f)$, and $\mathfrak{a}(c)$, we shall frequently write $R^{\mathfrak{A}}$, $f^{\mathfrak{A}}$, and $c^{\mathfrak{A}}$, or simply R^A, f^A, and c^A. For structures $\mathfrak{A}, \mathfrak{B}, \ldots$ we shall use $A, B, \ldots$ to denote their domains. Instead of writing an S-structure in the form $\mathfrak{A} = (A, \mathfrak{a})$, we shall often replace $\mathfrak{a}$ by a list of its values. For example, we write an $\{R, f, g\}$-structure as $\mathfrak{A} = (A, R^{\mathfrak{A}}, f^{\mathfrak{A}}, g^{\mathfrak{A}})$.

In investigations of arithmetic the symbol sets

$$S_{\text{ar}} := \{+, \cdot, 0, 1\} \qquad \text{and} \qquad S_{\text{ar}}^{<} := \{+, \cdot, 0, 1, <\}$$

play a special rôle, where $+$ and $\cdot$ are binary function symbols, 0 and 1 are constants, and $<$ is a binary relation symbol. Henceforth we shall use $\mathfrak{N}$ to denote the S_{ar}-structure $(\mathbb{N}, +^{\mathbb{N}}, \cdot^{\mathbb{N}}, 0^{\mathbb{N}}, 1^{\mathbb{N}})$, where $+^{\mathbb{N}}$ and $\cdot^{\mathbb{N}}$ are the usual addition and multiplication on $\mathbb{N}$ and $0^{\mathbb{N}}$ and $1^{\mathbb{N}}$ are the numbers zero and one, respectively. $\mathfrak{N}^{<} := (\mathbb{N}, +^{\mathbb{N}}, \cdot^{\mathbb{N}}, 0^{\mathbb{N}}, 1^{\mathbb{N}}, <^{\mathbb{N}})$, where $<^{\mathbb{N}}$ denotes the usual ordering on $\mathbb{N}$, is an example of an $S_{\text{ar}}^{<}$-structure. Similarly we set $\mathfrak{R} := (\mathbb{R}, +^{\mathbb{R}}, \cdot^{\mathbb{R}}, 0^{\mathbb{R}}, 1^{\mathbb{R}})$ and $\mathfrak{R}^{<} := (\mathbb{R}, +^{\mathbb{R}}, \cdot^{\mathbb{R}}, 0^{\mathbb{R}}, 1^{\mathbb{R}}, <^{\mathbb{R}})$. We shall often omit the superscripts $^{\mathbb{N}}, ^{\mathbb{R}}$ from $+^{\mathbb{N}}, +^{\mathbb{R}}, \ldots, 1^{\mathbb{N}}, 1^{\mathbb{R}}$. It will, however, be clear from the context whether, for example, $+$ is intended to denote the function symbol, the addition on $\mathbb{N}$, or the addition on $\mathbb{R}$.

The interpretation of variables is given by a so-called assignment.

1.2 Definition. An *assignment* in an S-structure $\mathfrak{A}$ is a map $\beta\colon \{v_n \mid n \in \mathbb{N}\} \to A$ of the set of variables into the domain A.

Now we can give a precise definition of the notion of interpretation:

1.3 Definition. An *S-interpretation* $\mathfrak{I}$ is a pair $(\mathfrak{A}, \beta)$ consisting of an S-structure $\mathfrak{A}$ and an assignment β in $\mathfrak{A}$.

When the particular symbol set S in question is either clear or unimportant we shall speak simply of structures and interpretations instead of S-structures and S-interpretations.

If β is an assignment in $\mathfrak{A}$, $a \in A$, and x is a variable, then let $\beta\frac{a}{x}$ be the assignment in $\mathfrak{A}$ which maps x to a and agrees with β on all variables distinct from x:

$$\beta\frac{a}{x}(y) = \begin{cases} \beta(y) & \text{if } y \neq x, \\ a & \text{if } y = x. \end{cases}$$

If $\mathfrak{I} = (\mathfrak{A}, \beta)$ let $\mathfrak{I}\frac{a}{x} = \left(\mathfrak{A}, \beta\frac{a}{x}\right)$.

In the introduction to this chapter we gave some examples showing how an S-formula can be read in everyday language once an S-interpretation has been given. It is useful to practice reading formulas under interpretations.

For example, if $S = S_{\mathrm{ar}}^{<}$, and the interpretation $\mathfrak{J} = (\mathfrak{A}, \beta)$ is given by

$$(*) \qquad \mathfrak{A} = (\mathbb{N}, +, \cdot, 0, 1, <) \qquad \text{and} \qquad \beta(v_n) = 2n \quad \text{for } n \geq 0,$$

then the formula $v_2 \cdot (v_1 + v_2) \equiv v_4$ (actually $\cdot v_2 + v_1 v_2 \equiv v_4$) reads "$4 \cdot (2 + 4) = 8$", and the formula $\forall v_0 \exists v_1 v_0 < v_1$ (actually $\forall v_0 \exists v_1 < v_0 v_1$) reads "for every natural number there is a larger natural number".

1.4 Exercise. Let $\mathfrak{J}$ be the interpretation defined above in (∗). How do the following formulas read with this interpretation?

$$\exists v_0 v_0 + v_0 \equiv v_1; \qquad \exists v_0 v_0 \cdot v_0 \equiv v_1; \qquad \exists v_1 v_0 \equiv v_1;$$

$$\forall v_0 \exists v_1 v_0 \equiv v_1; \qquad \forall v_0 \forall v_1 \exists v_2 (v_0 < v_2 \wedge v_2 < v_1).$$

1.5 Exercise. Let A be a finite nonempty set and S a finite symbol set. Show that there are only finitely many S-structures with A as domain.

1.6 Exercise. For S-structures $\mathfrak{A} = (A, \mathfrak{a})$ and $\mathfrak{B} = (B, \mathfrak{b})$, let $\mathfrak{A} \times \mathfrak{B}$ be the S-structure with domain $A \times B := \{(a, b) \mid a \in A, b \in B\}$, which is determined by the following conditions:

for n-ary R in S and $(a_0, b_0), \ldots, (a_{n-1}, b_{n-1}) \in A \times B$,

$$R^{\mathfrak{A} \times \mathfrak{B}}(a_0, b_0) \ldots (a_{n-1}, b_{n-1}) \quad \text{iff } (R^{\mathfrak{A}} a_0 \ldots a_{n-1} \text{ and } R^{\mathfrak{B}} b_0 \ldots b_{n-1});$$

for n-ary f in S and $(a_0, b_0), \ldots, (a_{n-1}, b_{n-1}) \in A \times B$,

$$f^{\mathfrak{A} \times \mathfrak{B}}((a_0, b_0), \ldots, (a_{n-1}, b_{n-1})) := (f^{\mathfrak{A}}(a_0, \ldots, a_{n-1}), f^{\mathfrak{B}}(b_0, \ldots, b_{n-1}));$$

for c in S,

$$c^{\mathfrak{A} \times \mathfrak{B}} := (c^{\mathfrak{A}}, c^{\mathfrak{B}}).$$

Show:

(a) If the S_{gr}-structures $\mathfrak{A}$ and $\mathfrak{B}$ are groups, then $\mathfrak{A} \times \mathfrak{B}$ is also a group.

(b) If $\mathfrak{A}$ and $\mathfrak{B}$ are equivalence structures, then $\mathfrak{A} \times \mathfrak{B}$ is also an equivalence structure.

(c) If the S_{ar}-structures $\mathfrak{A}$ and $\mathfrak{B}$ are fields, then $\mathfrak{A} \times \mathfrak{B}$ is not a field.

§2. Standardization of Connectives

When we define the notion of satisfaction in the next section we shall refer to the meaning of the connectives "not", "and", "or", "if-then", and "if and only if". In ordinary language their meanings vary. For example, "or" is sometimes used in an inclusive sense and at other times in the exclusive sense of "either-or". However, for our purposes it is useful to fix a standard meaning: We shall always use "or" in the inclusive sense, that is, a compound proposition whose constituents are connected by "or" is true (has the *truth-value* T) iff at least one of the constituents is true; it is false (has the *truth-value* F) iff both constituents are false. For example, we specify in 3.2 that a formula $(\varphi \vee \psi)$ is assigned the truth-value T under an interpretation $\mathfrak{I}$ if and only if φ is assigned the truth-value T under $\mathfrak{I}$ *or* ψ is assigned the truth-value T under $\mathfrak{I}$. On account of our fixed standard meaning we therefore have that $(\varphi \vee \psi)$ is assigned the truth-value T under $\mathfrak{I}$ if and only if at least one of the formulas φ, ψ is assigned T under $\mathfrak{I}$.

According to our convention the truth-value of a proposition compounded by "or" depends only on the truth-values of its constituents. Thus we can use a function

$$\dot{\vee} : \{T, F\} \times \{T, F\} \rightarrow \{T, F\}$$

to capture the meaning of "or"; the table of values ("truth-table") is as follows:

		$\dot{\vee}$
T	T	T
T	F	T
F	T	T
F	F	F

Similarly we proceed with the connectives "and", "if-then", "if and only if" and "not". The truth-tables for the functions $\dot{\wedge}$, $\dot{\rightarrow}$, $\dot{\leftrightarrow}$, and $\dot{\neg}$ are:

		$\dot{\wedge}$	$\dot{\rightarrow}$	$\dot{\leftrightarrow}$
T	T	T	T	T
T	F	F	F	F
F	T	F	T	F
F	F	F	T	T

	$\dot{\neg}$
T	F
F	T

These conventions correspond to mathematical practice.

Connectives for which the truth-values of compound propositions depend only on the truth-values of the constituents are called *extensional*. Thus we use the connectives "not", "and", "or", "if-then", and "if and only if" extensionally. In colloquial speech, however, these connectives are often not used extensionally. Consider, for example, "John fell ill and the doctor gave him a prescription", and "The doctor gave John a prescription and he fell ill". By contrast with the extensional case, the truth-value of these compound statements also depends on the temporal relation expressed by the order of the two components (*intensional* usage).

When we restrict ourselves to using the connectives extensionally, we sacrifice certain expressive possibilities of informal language. Experience shows, however, that this restriction is unimportant as far as the formulation of *mathematical* assertions is concerned. Furthermore, we shall see in the following exercise that all other extensional connectives can be defined from the connectives we have chosen.

2.1 Exercise. For $n \geq 1$ let $\mathbb{B}_n$ be the alphabet $\{\neg, \wedge, \vee, \rightarrow, \leftrightarrow,), (\} \cup \{p_0, \ldots, p_{n-1}\}$. We define the *formulas of the language of propositional calculus* (with the *propositional variables* $p_0, \ldots, p_{n-1}$) to be the strings α over $\mathbb{B}_n$ which can be obtained by means of the following rules:

$$\frac{}{p_i}\ (i < n), \qquad \frac{\alpha}{\neg\alpha} \quad \text{and} \quad \frac{\alpha,\ \beta}{(\alpha * \beta)} \quad \text{for } * = \wedge, \vee, \rightarrow, \leftrightarrow.$$

For an n-tuple $s = (s_0, \ldots, s_{n-1})$ of truth-values T, F, a so-called *assignment*, let $\alpha[s] \in \{\mathrm{T}, \mathrm{F}\}$ be defined by induction as follows:

$$p_i[s] = s_i \qquad (i < n);$$

$$\neg\alpha[s] = \dot{\neg}(\alpha[s]);$$

$$(\alpha * \beta)[s] = \dot{*}(\alpha[s], \beta[s]) \quad \text{for } * = \wedge, \vee, \rightarrow, \leftrightarrow.$$

$\alpha[s]$ is called the *truth-value* of α *with respect to the assignment* s.

(a) Show that for every $s = (s_0, s_1) \in \{\mathrm{T}, \mathrm{F}\}^2$,

$$((p_0 \wedge p_1) \vee (p_0 \wedge \neg p_1))[s] = s_0.$$

(b) An n-ary *truth-function* is a map $f\colon \{\mathrm{T}, \mathrm{F}\}^n \to \{\mathrm{T}, \mathrm{F}\}$. Show that for every $n \geq 1$ there are exactly 2^{2^n} n-ary truth-functions.

(c) Show that for every n-ary truth-function f there is a formula α in the propositional calculus in which the symbols $\wedge$, $\rightarrow$, and $\leftrightarrow$ do not occur, such that for all $s \in \{\mathrm{T}, \mathrm{F}\}^n$, $f(s) = \alpha[s]$. Prove the corresponding result for $\vee$, $\rightarrow$, and $\leftrightarrow$ instead of $\wedge$, $\rightarrow$, and $\leftrightarrow$.

(d) Analogously, one defines the language of propositional calculus with propositional variables $p_0, p_1, p_2, \ldots$, and considers corresponding assignments $s = (s_0, s_1, s_2, \ldots)$ of the truth-values T and F. Show that if at most the variables $p_0, p_1, \ldots, p_{n-1}$ occur in a formula α of this language, then $\alpha[s]$ depends only on $s_0, \ldots, s_{n-1}$.

§3. The Satisfaction Relation

The satisfaction relation makes precise the notion of a formula being true under an interpretation. Again we fix a symbol set S. By "term", "formula", or "interpretation" we always mean "S-term", "S-formula", or "S-interpretation". As a preliminary step we associate with every interpretation $\mathfrak{I} = (\mathfrak{A}, \beta)$ and every term t an element $\mathfrak{I}(t)$ from the domain A:

3.1 Definition. (a) For a variable x let $\mathfrak{I}(x) := \beta(x)$.
(b) For a constant $c \in S$, let $\mathfrak{I}(c) := c^{\mathfrak{A}}$.
(c) For an n-ary function symbol $f \in S$ and terms $t_0, \ldots, t_{n-1}$ let

$$\mathfrak{I}(ft_0 \ldots t_{n-1}) := f^{\mathfrak{A}}(\mathfrak{I}(t_0), \ldots, \mathfrak{I}(t_{n-1})).$$

As an illustration, if $S = S_{\text{gr}}$ and $\mathfrak{I} = (\mathfrak{R}_0, \beta)$ with $\mathfrak{R}_0 = (\mathbb{R}, +, 0)$ and $\beta(v_0) = 2$ and $\beta(v_2) = 6$, then $\mathfrak{I}(v_0 \circ (e \circ v_2)) = \mathfrak{I}(v_0) + \mathfrak{I}(e \circ v_2) = 2 + (0 + 6) = 8$.

Now, using induction on formulas φ, we give a definition of the relation $\mathfrak{I}$ *is a model of* φ, where $\mathfrak{I}$ is an arbitrary interpretation. If $\mathfrak{I}$ is a model of φ we also say that $\mathfrak{I}$ *satisfies* φ or that φ *holds in* $\mathfrak{I}$, and we write $\mathfrak{I} \models \varphi$.

3.2 Definition of the Satisfaction Relation. For all $\mathfrak{I} = (\mathfrak{A}, \beta)$ we let

$\mathfrak{I} \models t_0 \equiv t_1$ iff $\mathfrak{I}(t_0) = \mathfrak{I}(t_1)$,

$\mathfrak{I} \models Rt_0 \ldots t_{n-1}$ iff $R^{\mathfrak{A}}\mathfrak{I}(t_0) \ldots \mathfrak{I}(t_{n-1})$ (i.e., $R^{\mathfrak{A}}$ holds for $\mathfrak{I}(t_0), \ldots, \mathfrak{I}(t_{n-1})$).

$\mathfrak{I} \models \neg\varphi$ iff not $\mathfrak{I} \models \varphi$,

$\mathfrak{I} \models (\varphi \wedge \psi)$ iff $\mathfrak{I} \models \varphi$ and $\mathfrak{I} \models \psi$,

$\mathfrak{I} \models (\varphi \vee \psi)$ iff $\mathfrak{I} \models \varphi$ or $\mathfrak{I} \models \psi$,

$\mathfrak{I} \models (\varphi \rightarrow \psi)$ iff if $\mathfrak{I} \models \varphi$ then $\mathfrak{I} \models \psi$,

$\mathfrak{I} \models (\varphi \leftrightarrow \psi)$ iff $\mathfrak{I} \models \varphi$ if and only if $\mathfrak{I} \models \psi$,

$\mathfrak{I} \models \forall x\varphi$ iff for all $a \in A$ $\mathfrak{I}\frac{a}{x} \models \varphi$,

$\mathfrak{I} \models \exists x\varphi$ iff there is an $a \in A$ such that $\mathfrak{I}\frac{a}{x} \models \varphi$.

(For the definition of $\mathfrak{I}\frac{a}{x}$ see §1.)

Given a set Φ of S-formulas, we say that $\mathfrak{I}$ is a model of Φ and write $\mathfrak{I} \models \Phi$ if $\mathfrak{I} \models \varphi$ holds for all $\varphi \in \Phi$.

By going through the individual steps of definition 3.2 the reader should convince himself that $\mathfrak{I} \models \varphi$ if φ becomes a true statement under the interpretation $\mathfrak{I}$. The steps in 3.2 involving quantifiers are illustrated by the following example. Again let $S = S_{\text{gr}}$ and $\mathfrak{I} = (\mathfrak{R}_0, \beta)$ with $\mathfrak{R}_0 = (\mathbb{R}, +, 0)$ and $\beta(x) = 9$ for all x. Then we have

$$\mathfrak{I} \models \forall v_0 v_0 \circ e \equiv v_0 \quad \text{iff for all } r \in \mathbb{R}, \mathfrak{I}\frac{r}{v_0} \models v_0 \circ e \equiv v_0,$$
$$\text{iff for all } r \in \mathbb{R}, r + 0 = r.$$

3.3 Exercise. Let P be a unary relation symbol and f be a binary function symbol. For each of the formulas $\forall v_1 f v_0 v_1 \equiv v_0$, $\exists v_0 \forall v_1 f v_0 v_1 \equiv v_1$, and $\exists v_0(Pv_0 \wedge \forall v_1 Pfv_0 v_1)$ find an interpretation which satisfies the formula and one which does not satisfy it.

3.4 Exercise. A formula which does not contain $\neg$, $\leftrightarrow$, or $\rightarrow$ is called *positive*. Show that for every positive S-formula there is an S-interpretation which satisfies it. (*Hint*: One can, for example, use a one-element domain.)

§4. The Consequence Relation

Using the notion of satisfaction we can state exactly when a formula is a consequence of a set of formulas. Again we assume a symbol set S is given.

4.1 Definition of the Consequence Relation. We say that φ *is a consequence of* Φ (written: $\Phi \models \varphi$) iff every interpretation which is a model of Φ is also a model of φ.[1]

Instead of "$\{\psi\} \models \varphi$" we shall write "$\psi \models \varphi$". We have already sketched some examples of the consequence relation in Chapter I. Now we can formulate I.1.1. (existence of a left inverse) as follows:

$$\Phi_{\text{gr}} \models \forall v_0 \exists v_1 v_1 \circ v_0 \equiv e,$$

where

$$\Phi_{\text{gr}} = \{\forall v_0 \forall v_1 \forall v_2 v_0 \circ (v_1 \circ v_2) \equiv (v_0 \circ v_1) \circ v_2, \\ \forall v_0 v_0 \circ e \equiv v_0, \forall v_0 \exists v_1 v_0 \circ v_1 \equiv e\}.$$

To show that a formula φ is *not* a consequence of a set of formulas Φ, it is sufficient to give an interpretation which satisfies every formula in Φ but fails to satisfy φ.

For example, one shows

$$\text{(1)} \qquad \text{not } \Phi_{\text{gr}} \models \forall v_0 \forall v_1 v_0 \circ v_1 \equiv v_1 \circ v_0$$

by giving as an interpretation a nonabelian group $\mathfrak{G}$ with an arbitrary assignment of variables to elements of $\mathfrak{G}$. Analogously, one can use an abelian group to show

$$\text{(2)} \qquad \text{not } \Phi_{\text{gr}} \models \neg \forall v_0 \forall v_1 v_0 \circ v_1 \equiv v_1 \circ v_0.$$

[1] We use the symbol $\models$ for both the satisfaction relation ($\mathfrak{I} \models \varphi$) and for the consequence relation ($\Phi \models \varphi$). The symbol preceding "$\models$" (either for an interpretation, such as $\mathfrak{I}$, or for a set of formulas, such as Φ) determines the meaning.

With (1) and (2) we see that

$$\text{not } \Phi \models \varphi$$

does not necessarily imply

$$\Phi \models \neg\varphi.$$

In Chapter I it became clear—both by examples and in an informal way—that when φ can be proved from a system of axioms Φ then φ is a consequence of Φ. There we raised the question as to what extent the consequences of a system of axioms can be obtained by mathematical proofs. The precise definitions of concepts given in this chapter and the next lay the foundation for a rigorous discussion of this question. In Chapter V we shall obtain the fundamental result that the consequence relation $\Phi \models \varphi$ can always be established by means of a mathematical proof. We shall see that such a proof consists of very elementary steps which, moreover, can be described in a purely formal way (that is, syntactically).

Using the notion of consequence we are now able to define the notions of *validity*, *satisfiability*, and *logical equivalence.*

4.2 Definition. A formula φ is *valid* (written: $\models \varphi$) iff $\emptyset \models \varphi$.

Thus a formula is valid when it holds under all interpretations. For example, all formulas of the form $(\varphi \vee \neg\varphi)$ or $\exists x\, x \equiv x$ are valid.

4.3 Definition. A formula φ is *satisfiable* (written: Sat φ) iff there is an interpretation which is a model of φ. A set of formulas Φ is satisfiable (written: Sat Φ) iff there is an interpretation which is a model of all the formulas in Φ.

4.4 Lemma. *For all* Φ *and all* φ,

$$\Phi \models \varphi \quad \textit{iff not } \text{Sat } \Phi \cup \{\neg\varphi\}.$$

In particular, φ *is valid iff* $\neg\varphi$ *is not satisfiable.*

PROOF. $\Phi \models \varphi$

iff every interpretation which is a model of Φ is also a model of φ,

iff there is no interpretation which is a model of Φ but not a model of φ,

iff there is no interpretation which is a model of $\Phi \cup \{\neg\varphi\}$,

iff not Sat $\Phi \cup \{\neg\varphi\}$. □

4.5 Definition. Two formulas φ and ψ are *logically equivalent* iff $\varphi \models \psi$ and $\psi \models \varphi$.

Thus two formulas φ and ψ are logically equivalent iff they are valid under the same interpretations, that is, iff $\models \varphi \leftrightarrow \psi$.

It is immediately evident, from the definition of the notion of satisfaction together with the truth-tables for connectives that the following formulas are logically equivalent:

$$(+)\qquad \begin{array}{lll} \varphi \wedge \psi & \text{and} & \neg(\neg\varphi \vee \neg\psi), \\ \varphi \to \psi & \text{and} & \neg\varphi \vee \psi, \\ \varphi \leftrightarrow \psi & \text{and} & \neg(\varphi \vee \psi) \vee \neg(\neg\varphi \vee \neg\psi), \\ \forall x\varphi & \text{and} & \neg\exists x \neg \varphi. \end{array}$$

Therefore, we can dispense with the connectives $\wedge$, $\to$, and $\leftrightarrow$, and the quantifier $\forall$. More precisely, we define a map * by induction on formulas, which associates with every formula φ a formula φ^* such that φ^* is logically equivalent to φ and does not contain $\wedge$, $\to$, $\leftrightarrow$, or $\forall$:

$$\begin{aligned} \varphi^* &:= \varphi \quad \text{if } \varphi \text{ is atomic}, \\ (\neg\varphi)^* &:= \neg\varphi^*, \\ (\varphi \vee \psi)^* &:= \varphi^* \vee \psi^*, \\ (\varphi \wedge \psi)^* &:= \neg(\neg\varphi^* \vee \neg\psi^*), \\ (\varphi \to \psi)^* &:= \neg\varphi^* \vee \psi^*, \\ (\varphi \leftrightarrow \psi)^* &:= \neg(\varphi^* \vee \psi^*) \vee \neg(\neg\varphi^* \vee \neg\psi^*), \\ (\exists x\varphi)^* &:= \exists x\varphi^*, \\ (\forall x\varphi)^* &:= \neg\exists x \neg\varphi^*. \end{aligned}$$

Using (+) one can easily prove that * has the desired properties.

In general a formula φ is easier to read than the corresponding formula φ^*, as is clear from (+). But because of the logical equivalence of φ and φ^* we do not lose expressive power when we exclude the symbols $\wedge$, $\to$, $\leftrightarrow$, and $\forall$ from our first-order languages. This will simplify our investigations of the languages; in particular, proofs by induction on formulas will be shorter. Thus we make the following conventions:

(1) *In the sequel we restrict ourselves to formulas in which* $\wedge$, $\to$, $\leftrightarrow$, *and* $\forall$ *do not occur*. That is, in the common alphabet $\mathbb{A}$ of the first-order languages (cf. II.2.1.) we omit the symbols $\wedge$, $\to$, $\leftrightarrow$, and $\forall$. In the definition II.3.2. of formulas we restrict cases (F4) and (F5) to the introduction of formulas of the form $(\varphi \vee \psi)$ and $\exists x\varphi$, respectively. Finally, in the definition of the notion of satisfaction we eliminate the cases corresponding to $\wedge$, $\to$, $\leftrightarrow$, and $\forall$.

(2) Nevertheless we shall sometimes retain the symbols $\wedge$, $\to$, $\leftrightarrow$, $\forall$ when writing formulas. Such "formulas φ in the old style" should now be understood as abbreviations for φ^*; for example, $\forall x(Px \wedge Qx)$ should be understood as an abbreviation for $\neg\exists x \neg\neg(\neg Px \vee \neg Qx)$.

4.6 Exercise. For arbitrary formulas φ, ψ, and χ show that:

(a) $(\varphi \vee \psi) \models \chi$ iff $\varphi \models \chi$ and $\psi \models \chi$.
(b) $\models (\varphi \rightarrow \psi)$ iff $\varphi \models \psi$.

4.7 Exercise. (a) Show that $\exists x \, \forall y \varphi \models \forall y \, \exists x \varphi$.
(b) Show that $\forall y \, \exists x \, Rxy \models \exists x \, \forall y \, Rxy$ does not hold.

4.8 Exercise. Prove:

(a) $\models \forall x(\varphi \wedge \psi) \leftrightarrow \forall x \varphi \wedge \forall x \psi$,
(b) $\models \exists x(\varphi \vee \psi) \leftrightarrow \exists x \varphi \vee \exists x \psi$,
(c) $\models \forall x(\varphi \vee \psi) \leftrightarrow \varphi \vee \forall x \psi$, if $x \notin \text{free}(\varphi)$,
(d) $\models \exists x(\varphi \wedge \psi) \leftrightarrow \varphi \wedge \exists x \psi$, if $x \notin \text{free}(\varphi)$.
(e) Show that one cannot do without the assumption "$x \notin \text{free}(\varphi)$" in (c) and (d).

4.9 Exercise. Let φ and ψ be formulas such that $\models (\varphi \leftrightarrow \psi)$. Let χ' be obtained from the formula χ by replacing all subformulas of the form φ by ψ.

(a) Define the map $'$ by induction on formulas.
(b) Show that for all χ, $\models (\chi \leftrightarrow \chi')$.

Historical Note. The precise development of semantics is due essentially to A. Tarski [27]. The notion of logical consequence was already present in work of B. Bolzano [3].

§5. Coincidence Lemma and Isomorphism Lemma

It seems intuitively clear that the satisfaction relation between an S-formula φ and an S-interpretation $\mathfrak{I}$ depends only on the interpretation of the *symbols of S occurring in* φ, and on the *variables occurring free in* φ. The following lemma gives an exact formulation.

5.1 Coincidence Lemma. *Let* $\mathfrak{I}_1 = (\mathfrak{A}_1, \beta_1)$ *be an* S_1*-interpretation and* $\mathfrak{I}_2 = (\mathfrak{A}_2, \beta_2)$ *be an* S_2*-interpretation, both with the same domain,* i.e., $A_1 = A_2$. *Put* $S := S_1 \cap S_2$.

(a) *Let t be an S-term. If* $\mathfrak{I}_1$ *and* $\mathfrak{I}_2$ *agree*[2] *on the S-symbols occurring in t and on the variables occurring in t, then* $\mathfrak{I}_1(t) = \mathfrak{I}_2(t)$.
(b) *Let* φ *be an S-formula. If* $\mathfrak{I}_1$ *and* $\mathfrak{I}_2$ *agree on the S-symbols and on the variables occurring free in* φ, *then* $\mathfrak{I}_1 \models \varphi$ *iff* $\mathfrak{I}_2 \models \varphi$.

[2] $\mathfrak{I}_1$ and $\mathfrak{I}_2$ agree on $k \in S$ or on x if $k^{\mathfrak{A}_1} = k^{\mathfrak{A}_2}$ or $\beta_1(x) = \beta_2(x)$, respectively.

Proof. (a) We use induction on terms.

$t = x$: By hypothesis, $\beta_1(x) = \beta_2(x)$ and therefore

$$\mathfrak{I}_1(x) = \beta_1(x) = \beta_2(x) = \mathfrak{I}_2(x).$$

$t = c$: Similarly.

$t = ft_0 \ldots t_{n-1}$ ($f \in S$ n-ary and $t_0, \ldots, t_{n-1} \in T^S$):

$$\begin{aligned}\mathfrak{I}_1(ft_0 \ldots t_{n-1}) &= f^{\mathfrak{A}_1}(\mathfrak{I}_1(t_0), \ldots, \mathfrak{I}_1(t_{n-1}))\\ &= f^{\mathfrak{A}_1}(\mathfrak{I}_2(t_0), \ldots, \mathfrak{I}_2(t_{n-1})) \quad \text{(by induction hypothesis)}\\ &= f^{\mathfrak{A}_2}(\mathfrak{I}_2(t_0), \ldots, \mathfrak{I}_2(t_{n-1})) \quad \text{(by hypothesis } f^{\mathfrak{A}_1} = f^{\mathfrak{A}_2}\text{)}\\ &= \mathfrak{I}_2(ft_0 \ldots t_{n-1}).\end{aligned}$$

(b) We use induction on S-formulas.

$\varphi = Rt_0 \ldots t_{n-1}$ ($R \in S$ n-ary and $t_0, \ldots, t_{n-1} \in T^S$):

$$\begin{aligned}\mathfrak{I}_1 \models Rt_0 \ldots t_{n-1} \quad &\text{iff } R^{\mathfrak{A}_1}\mathfrak{I}_1(t_0) \ldots \mathfrak{I}_1(t_{n-1})\\ &\text{iff } R^{\mathfrak{A}_1}\mathfrak{I}_2(t_0) \ldots \mathfrak{I}_2(t_{n-1}) \quad \text{(by (a))}\\ &\text{iff } R^{\mathfrak{A}_2}\mathfrak{I}_2(t_0) \ldots \mathfrak{I}_2(t_{n-1}) \quad \text{(by hypothesis } R^{\mathfrak{A}_1} = R^{\mathfrak{A}_2}\text{)}\\ &\text{iff } \mathfrak{I}_2 \models Rt_0 \ldots t_{n-1}.\end{aligned}$$

$\varphi = t_1 \equiv t_2$: Similarly.

$\varphi = \neg\psi$:

$$\begin{aligned}\mathfrak{I}_1 \models \neg\psi \quad &\text{iff not } \mathfrak{I}_1 \models \psi\\ &\text{iff not } \mathfrak{I}_2 \models \psi \quad \text{(by induction hypothesis)}\\ &\text{iff } \mathfrak{I}_2 \models \neg\psi.\end{aligned}$$

$\varphi = (\psi \vee \chi)$: Similarly.

$\varphi = \exists x\psi$:

$$\begin{aligned}\mathfrak{I}_1 \models \exists x\psi \quad &\text{iff there is an } a \in A_1 \text{ such that } \mathfrak{I}_1\frac{a}{x} \models \psi\\ &\text{iff there is an } a \in A_2 (= A_1) \text{ such that } \mathfrak{I}_2\frac{a}{x} \models \psi\end{aligned}$$

(by the induction hypothesis applied to ψ, $\mathfrak{I}_1\frac{a}{x}$ and $\mathfrak{I}_2\frac{a}{x}$; note that, because free(ψ) $\subset$ free(φ) $\cup \{x\}$, the interpretations $\mathfrak{I}_1\frac{a}{x}$ and $\mathfrak{I}_2\frac{a}{x}$ agree on all symbols occurring in ψ and all variables occurring free in ψ)

$$\text{iff } \mathfrak{I}_2 \models \exists x\psi.$$

□

In particular, the coincidence lemma says that, for an S-formula φ and an S-interpretation $\mathfrak{I} = (\mathfrak{A}, \beta)$, the validity of φ under $\mathfrak{I}$ depends only on the assignments for the *finitely* many variables occurring free in φ (and, of course, on the interpretation of the symbols of S in $\mathfrak{A}$). If these variables are among $v_0, \ldots, v_{n-1}$, i.e., if $\varphi \in L_n^S$, it is at most the β-values $a_i = \beta(v_i)$ for $i = 0, \ldots, n-1$ which are significant. Thus, instead of $(\mathfrak{A}, \beta) \models \varphi$, we shall often use the more suggestive notation

$$\mathfrak{A} \models \varphi[a_0, \ldots, a_{n-1}].$$

(Similarly, for an S-term t such that $\mathrm{var}(t) \subset \{v_0, \ldots, v_{n-1}\}$ we shall write $t^{\mathfrak{A}}[a_0, \ldots, a_{n-1}]$ instead of $\mathfrak{I}(t)$.) When φ is a sentence, i.e., $\varphi \in L_0^S$, we can choose $n = 0$ and write

$$\mathfrak{A} \models \varphi,$$

without even mentioning an assignment. In that case we say that $\mathfrak{A}$ *is a model of* φ. For a set of sentences Φ, $\mathfrak{A} \models \Phi$ means that $\mathfrak{A} \models \varphi$ for every $\varphi \in \Phi$.

5.2 Definition. Let S and S' be symbol sets such that $S \subset S'$; let $\mathfrak{A} = (A, \mathfrak{a})$ be an S-structure, and $\mathfrak{A}' = (A', \mathfrak{a}')$ be an S'-structure. We call $\mathfrak{A}$ a *reduct* of $\mathfrak{A}'$ (more precisely, the S-*reduct* of $\mathfrak{A}'$) iff $A = A'$ and $\mathfrak{a}$ and $\mathfrak{a}'$ agree on S. In this case $\mathfrak{A}'$ is called an *expansion* of $\mathfrak{A}$, and we write $\mathfrak{A} = \mathfrak{A}' \upharpoonright S$.

For example, the $S_{\mathrm{ar}}^{<}$-structure $\mathfrak{R}^{<}$ (the ordered field of real numbers) is an expansion of the S_{ar}-structure $\mathfrak{R}$ (the field of real numbers): $\mathfrak{R} = \mathfrak{R}^{<} \upharpoonright S_{\mathrm{ar}}$.

If $\mathfrak{A} = \mathfrak{A}' \upharpoonright S$, then it follows from the coincidence lemma that for $\varphi \in L_n^S$ and $a_0, \ldots, a_{n-1} \in A$, $\mathfrak{A} \models \varphi[a_0, \ldots, a_{n-1}]$ iff $\mathfrak{A}' \models \varphi[a_0, \ldots, a_{n-1}]$. To see that this holds we choose $\beta\colon \{v_n : n \in \mathbb{N}\} \to A$ so that $\beta(v_i) = a_i$ for $i < n$, and we apply the coincidence lemma for $\mathfrak{I}_1 = (\mathfrak{A}, \beta)$ and $\mathfrak{I}_2 = (\mathfrak{A}', \beta)$; $\mathfrak{I}_1$ and $\mathfrak{I}_2$ agree on the symbols occurring in φ and on the variables occurring free in φ.

The definitions of interpretation, consequence, and satisfiability refer to a fixed symbol set S. Using the coincidence lemma we can remove this reference to S. Let us consider, for example, the notion of satisfiability. If Φ is a set of S-formulas and $S' \supset S$, then Φ is also a set of S'-formulas. As a set of S-formulas, Φ is satisfiable if there is an S-interpretation which satisfies it, and as a set of S'-formulas it is satisfiable if there is an S'-interpretation which satisfies it. We have

5.3 Lemma. *Φ is satisfiable with respect to S iff Φ is satisfiable with respect to S'.*

Proof. If $\mathfrak{I}' = (\mathfrak{A}', \beta')$ is an S'-interpretation such that $\mathfrak{I}' \models \Phi$, then by 5.1 the S-interpretation $(\mathfrak{A}' \upharpoonright S, \beta')$ is a model of Φ. On the other hand, if $\mathfrak{I} = (\mathfrak{A}, \beta)$ is an S-interpretation which satisfies Φ, we choose an S'-structure $\mathfrak{A}'$ such that $\mathfrak{A}' \upharpoonright S = \mathfrak{A}$. (The symbols in $S' - S$ can be interpreted arbitrarily.) By 5.1 the S-interpretation $(\mathfrak{A}', \beta)$ is then a model of Φ. □

We conclude this section with a result about isomorphic structures.

5.4 Definition. Let $\mathfrak{A}$ and $\mathfrak{B}$ be S-structures.

(a) A map $\pi: A \to B$ is called an *isomorphism of* $\mathfrak{A}$ *onto* $\mathfrak{B}$ (written: $\pi: \mathfrak{A} \cong \mathfrak{B}$) iff

(1) π is a bijection of A onto B;

(2) for n-ary $R \in S$ and $a_0, \ldots, a_{n-1} \in A$,

$$R^{\mathfrak{A}} a_0 \ldots a_{n-1} \quad \text{iff } R^{\mathfrak{B}} \pi(a_0) \ldots \pi(a_{n-1});$$

(3) for n-ary $f \in S$ and $a_0, \ldots, a_{n-1} \in A$,

$$\pi(f^{\mathfrak{A}}(a_0, \ldots, a_{n-1})) = f^{\mathfrak{B}}(\pi(a_0), \ldots, \pi(a_{n-1}));$$

(4) for every $c \in S$, $\pi(c^{\mathfrak{A}}) = c^{\mathfrak{B}}$.

(b) $\mathfrak{A}$ and $\mathfrak{B}$ are said to be *isomorphic* (written: $\mathfrak{A} \cong \mathfrak{B}$) iff there is an isomorphism $\pi: \mathfrak{A} \cong \mathfrak{B}$.

For example, the S_{gr}-structure $(\mathbb{N}, +, 0)$ is isomorphic to the S_{gr}-structure $(G, +^G, 0)$ consisting of the even natural numbers with ordinary addition $+^G$. The map $\pi: \mathbb{N} \to G$ such that $\pi(n) = 2n$ is an isomorphism of $(\mathbb{N}, +, 0)$ onto $(G, +^G, 0)$.

The following lemma shows that isomorphic structures cannot be distinguished by means of first-order sentences.

5.5 Isomorphism Lemma. *If* $\mathfrak{A}$ *and* $\mathfrak{B}$ *are isomorphic S-structures, then for all S-sentences* φ

$$\mathfrak{A} \models \varphi \quad \textit{iff } \mathfrak{B} \models \varphi.$$

Proof. Let $\pi: \mathfrak{A} \cong \mathfrak{B}$. For the intended proof by induction it is convenient to show not only that the same S-sentences hold in $\mathfrak{A}$ and $\mathfrak{B}$, but also that the same S-formulas hold if one uses corresponding assignments: With every assignment β in $\mathfrak{A}$ we associate the assignment $\beta^\pi := \pi \circ \beta$ in $\mathfrak{B}$, and for the corresponding interpretations $\mathfrak{I} = (\mathfrak{A}, \beta)$ and $\mathfrak{I}^\pi := (\mathfrak{B}, \beta^\pi)$ we shall show:

(i) For every S-term t, $\pi(\mathfrak{I}(t)) = \mathfrak{I}^\pi(t)$.

(ii) For every S-formula φ, $\mathfrak{I} \models \varphi$ iff $\mathfrak{I}^\pi \models \varphi$.

This will complete the proof.

(i) can easily be proved by induction on terms. (ii) is proved by induction on formulas φ simultaneously for all assignments β in $\mathfrak{A}$. We only treat the case of atomic formulas and the steps involving $\neg$ and $\exists$.

$\varphi = t_0 \equiv t_1$:

$$
\begin{aligned}
\mathfrak{I} \models t_0 \equiv t_1 \quad &\text{iff } \mathfrak{I}(t_0) = \mathfrak{I}(t_1) \\
&\text{iff } \pi(\mathfrak{I}(t_0)) = \pi(\mathfrak{I}(t_1) \quad (\text{since } \pi\colon A \to B \text{ is injective}) \\
&\text{iff } \mathfrak{I}^\pi(t_0) = \mathfrak{I}^\pi(t_1) \quad (\text{by (i)}) \\
&\text{iff } \mathfrak{I}^\pi \models t_0 \equiv t_1.
\end{aligned}
$$

$\varphi = Rt_0 \ldots t_{n-1}$:

$$
\begin{aligned}
\mathfrak{I} \models Rt_0 \ldots t_{n-1} \quad &\text{iff } R^{\mathfrak{A}}\mathfrak{I}(t_0) \ldots \mathfrak{I}(t_{n-1}) \\
&\text{iff } R^{\mathfrak{B}}\pi(\mathfrak{I}(t_0)) \ldots \pi(\mathfrak{I}(t_{n-1})) \quad (\text{because } \pi\colon \mathfrak{A} \cong \mathfrak{B}) \\
&\text{iff } R^{\mathfrak{B}}\mathfrak{I}^\pi(t_0) \ldots \mathfrak{I}^\pi(t_{n-1}) \quad (\text{by (i)}) \\
&\text{iff } \mathfrak{I}^\pi \models Rt_0 \ldots t_{n-1}.
\end{aligned}
$$

$\varphi = \neg\psi$:

$$
\begin{aligned}
\mathfrak{I} \models \neg\psi \quad &\text{iff not } \mathfrak{I} \models \psi \\
&\text{iff not } \mathfrak{I}^\pi \models \psi \quad (\text{by induction hypothesis}) \\
&\text{iff } \mathfrak{I}^\pi \models \neg\psi.
\end{aligned}
$$

$\varphi = \exists x\psi$:

$$
\begin{aligned}
\mathfrak{I} \models \exists x\psi \quad &\text{iff there is an } a \in A \text{ such that } \mathfrak{I}\frac{a}{x} \models \psi \\
&\text{iff there is an } a \in A \text{ such that } \left(\mathfrak{I}\frac{a}{x}\right)^\pi \models \psi \\
&\qquad (\text{by induction hypothesis}) \\
&\text{iff there is an } a \in A \text{ such that } \mathfrak{I}^\pi\frac{\pi(a)}{x} \models \psi \\
&\qquad \left(\text{since } \left(\mathfrak{I}\frac{a}{x}\right)^\pi = \mathfrak{I}^\pi\frac{\pi(a)}{x}\right) \\
&\text{iff there is an element } b \in B \text{ such that } \mathfrak{I}^\pi\frac{b}{x} \models \psi \\
&\qquad (\text{since } \pi\colon A \to B \text{ is surjective}) \\
&\text{iff } \mathfrak{I}^\pi \models \exists x\psi. \qquad \square
\end{aligned}
$$

From the proof we infer

5.6 Corollary. *If $\pi\colon \mathfrak{A} \cong \mathfrak{B}$, then for $\varphi \in L_n^S$ and $a_0, \ldots, a_{n-1} \in A$,*

$$\mathfrak{A} \models \varphi[a_0, \ldots, a_{n-1}] \quad \textit{iff } \mathfrak{B} \models \varphi[\pi(a_0), \ldots, \pi(a_{n-1})]. \qquad \square$$

5.5 tells us that isomorphic S-structures cannot be distinguished in L_0^S. Conversely one could ask whether S-structures in which the same S-sentences are satisfied are isomorphic. In Chapter VI we shall see that this is not always the case. For example, there are structures not isomorphic to the S_{ar}-structure $\mathfrak{N}$ of natural numbers in which the same first-order sentences hold.

5.7 Exercise. Prove the analogue of 5.3 for the consequence relation.

5.8 Exercise. Let S be a finite symbol set and let $\mathfrak{A}$ be a finite S-structure. Show that there is an S-sentence $\varphi_{\mathfrak{A}}$ the models of which are precisely the S-structures isomorphic to $\mathfrak{A}$.

5.9 Exercise. Show:

(a) The relation $<$ ("less-than") is elementarily definable in $(\mathbb{R}, +, \cdot, 0)$, i.e., there is a formula $\varphi \in L_2^{\{+, \cdot, 0\}}$ such that for all $a, b \in \mathbb{R}$, $(\mathbb{R}, +, \cdot, 0) \models \varphi[a, b]$ iff $a < b$.

(b) The relation $<$ is not elementarily definable in $(\mathbb{R}, +, 0)$. (*Hint*: Work with a suitable *automorphism* of $(\mathbb{R}, +, 0)$, i.e., a suitable isomorphism of $(\mathbb{R}, +, 0)$ onto itself.)

5.10 Exercise. Let I be a nonempty set. For every $i \in I$, let $\mathfrak{A}_i$ be an S-structure. We write $\prod_{i \in I} \mathfrak{A}_i$ for the *direct product* of the structures $\mathfrak{A}_i$, $i \in I$, that is, the S-structure $\mathfrak{A}$ with domain $\prod_{i \in I} A_i := \{g \mid g: I \to \bigcup_{i \in I} A_i$, and for all $i \in I$, $g(i) \in A_i\}$, which is determined by the following conditions (for $g \in \prod_{i \in I} A_i$ we also write $\langle g(i) \mid i \in I \rangle$):

For n-ary R in S and for $g_0, \ldots, g_{n-1} \in \prod_{i \in I} A_i$,

$$R^{\mathfrak{A}} g_0 \ldots g_{n-1} \quad \text{iff for all } i \in I,\ R^{\mathfrak{A}_i} g_0(i) \ldots g_{n-1}(i);$$

for n-ary f in S and $g_0, \ldots, g_{n-1} \in \prod_{i \in I} A_i$,

$$f^{\mathfrak{A}}(g_0, \ldots, g_{n-1}) := \langle f^{\mathfrak{A}_i}(g_0(i), \ldots, g_{n-1}(i)) \mid i \in I \rangle;$$

for c in S,

$$c^{\mathfrak{A}} := \langle c^{\mathfrak{A}_i} \mid i \in I \rangle.$$

(a) Show that if t is an S-term such that $\operatorname{var}(t) \subset \{v_0, \ldots, v_{n-1}\}$ and if $g_0, \ldots, g_{n-1} \in \prod_{i \in I} A_i$, then

$$t^{\mathfrak{A}}[g_0, \ldots, g_{n-1}] = \langle t^{\mathfrak{A}_i}[g_0(i), \ldots, g_{n-1}(i)] \mid i \in I \rangle.$$

Formulas which are derivable in the following calculus are called *Horn formulas*.

(i) $\dfrac{}{\varphi}$, if φ is atomic;

(ii) $\dfrac{}{\neg \varphi}$, if φ is atomic;

(iii) $\dfrac{}{(\varphi_0 \wedge \cdots \wedge \varphi_{n-1}) \to \varphi_n}$, if $\varphi_0, \ldots, \varphi_n$ are atomic and $n \geq 1$;

(iv) $\dfrac{\varphi, \quad \psi}{(\varphi \wedge \psi)}$;

(v) $\dfrac{\varphi}{\forall x \varphi}$;

(vi) $\dfrac{\varphi}{\exists x \varphi}$.

Horn sentences are Horn formulas containing no free variables.

(b) Show that if φ is a Horn sentence and if every $\mathfrak{A}_i$ is a model of φ, then $\prod_{i \in I} \mathfrak{A}_i \models \varphi$. (*Hint*: State and prove the corresponding result for Horn formulas.)

5.11 Exercise. A set Φ of sentences is called *independent* if there is no $\varphi \in \Phi$ such that $\Phi - \{\varphi\} \models \varphi$. Show that the axioms for groups (cf. p. 32) and the axioms for equivalence relations (cf. p. 16) are independent sets of sentences.

§6. Some Simple Formalizations

As we saw in §4, the axioms for group theory can be formulated, or as we often say, formalized, in a first-order language. As another example of formalization we give the cancellation law for group theory:

$$\varphi := \forall v_0 \, \forall v_1 \, \forall v_2 (v_0 \circ v_2 \equiv v_1 \circ v_2 \rightarrow v_0 \equiv v_1).$$

To say that the cancellation law holds in a group $\mathfrak{G}$ means that $\mathfrak{G} \models \varphi$, and to say that it holds in all groups means that $\Phi_{\mathrm{gr}} \models \varphi$.

The statement "there is no element of order two" can be formalized as

$$\psi := \neg \exists v_0 (\neg v_0 \equiv e \wedge v_0 \circ v_0 \equiv e).$$

The observation that there is no element of order two in $(\mathbb{Z}, +, 0)$ thus means that $(\mathbb{Z}, +, 0)$ is a model of ψ.

For applications of our results it is helpful to have a certain proficiency in formalization. The following examples should serve this purpose. As the exact choice of variables is unimportant (for example, instead of using the formula φ above we could have used

$$\forall v_{17} \, \forall v_8 \, \forall v_1 (v_{17} \circ v_1 \equiv v_8 \circ v_1 \rightarrow v_{17} \equiv v_8)$$

to formalize the cancellation law) we shall denote the variables simply by $x, y, z, \ldots$, where distinct letters stand for *distinct* variables.

6.1 Equivalence Relations. The three defining properties of an equivalence relation can be formalized with the aid of a single binary relation symbol as follows:

$$\forall x\, Rxx, \qquad \forall x\, \forall y(Rxy \rightarrow Ryx), \qquad \forall x\, \forall y\, \forall z((Rxy \wedge Ryz) \rightarrow Rxz).$$

The theorem mentioned in Chapter I

> If x and y are both equivalent to a third element then they are equivalent to the same elements,

can be reformulated as

> For all x and y, if there is an element u such that x is equivalent to u and y is equivalent to u, then for all z, x is equivalent to z iff y is equivalent to z,

and then formalized as

$$\forall x\, \forall y(\exists u(Rxu \wedge Ryu) \rightarrow \forall z(Rxz \leftrightarrow Ryz)).$$

6.2 Continuity. Let ρ be a unary function on $\mathbb{R}$ and let Δ be the binary distance function on $\mathbb{R}$, that is, $\Delta(r_0, r_1) = |r_0 - r_1|$ for $r_0, r_1 \in \mathbb{R}$. Using the function symbols f (for ρ) and d (for Δ) we can treat $(\mathbb{R}, +, \cdot, 0, 1, <, \rho, \Delta)$ as an $S_{ar}^{<} \cup \{f, d\}$-structure. The continuity of ρ on $\mathbb{R}$ can be stated as follows:

(*) For all x and for all $\varepsilon > 0$ there is a $\delta > 0$ such that for all y, if $\Delta(x, y) < \delta$ then $\Delta(\rho(x), \rho(y)) < \varepsilon$.

Concerning the restricted quantifiers "for all $\varepsilon > 0$" and "there is a $\delta > 0$" that appear in (*) it is useful to observe that a statement of the form

for all x such that ..., we have ——

can be formalized

$$\forall x(\ldots \rightarrow \text{——}),$$

and a statement of the type

there is an x with ... such that ——

can be formalized

$$\exists x(\ldots \wedge \text{——}).$$

Thus, using the variables u and v for ε and δ we can give the following formalization of (*):

$$\forall x\, \forall u(0 < u \rightarrow \exists v(0 < v \wedge \forall y(dxy < v \rightarrow dfxfy < u))).$$

6.3 Cardinality Statements. The sentence

$$\varphi_{\geq 2} := \exists v_0\, \exists v_1\, \neg v_0 \equiv v_1$$

is a formalization of "there are at least two elements". More precisely, for all S and all S-structures $\mathfrak{A}$,

$$\mathfrak{A} \models \varphi_{\geq 2} \quad \text{iff A contains at least two elements.}$$

In a similar way, for $n \geq 3$, the sentence

$$\varphi_{\geq n} := \exists v_0 \ldots \exists v_{n-1}(\neg v_0 \equiv v_1 \wedge \cdots \wedge \neg v_0 \equiv v_{n-1} \wedge \cdots \wedge \neg v_{n-2} \equiv v_{n-1})$$

states that there are at least n elements, and the sentences

$$\neg \varphi_{\geq n} \qquad \text{and} \qquad \varphi_{\geq n} \wedge \neg \varphi_{\geq n+1}$$

say that there are fewer than n elements and exactly n elements, respectively. If we now put

$$\Phi_\infty := \{\varphi_{\geq n} \mid n \geq 2\},$$

then the models of Φ_∞ are precisely the infinite structures, that is, for all S and all S-structures $\mathfrak{A}$,

$$\mathfrak{A} \models \Phi_\infty \quad \text{iff A contains infinitely many elements.}$$

6.4 The Theory of Orderings. A structure $\mathfrak{A} = (A, <^{\mathfrak{A}})$ is called an *ordering* if it is a model of the following sentences:

$$\Phi_{\text{ord}} \begin{cases} \forall x \, \neg x < x, \\ \forall x \, \forall y \, \forall z((x < y \wedge y < z) \rightarrow x < z), \\ \forall x \, \forall y(x < y \vee x \equiv y \vee y < x). \end{cases}$$

$(\mathbb{R}, <^{\mathbb{R}})$ and $(\mathbb{N}, <^{\mathbb{N}})$ are examples of orderings. If $\mathbb{C}$ denotes the set of complex numbers and $<^{\mathbb{C}}$ is defined by

$$z_1 <^{\mathbb{C}} z_2 \quad \text{iff } z_1, z_2 \in \mathbb{R} \text{ and } z_1 <^{\mathbb{R}} z_2$$

then $(\mathbb{C}, <^{\mathbb{C}})$ is not an ordering because the third axiom in Φ_{ord} is violated. If for a structure $\mathfrak{A} = (A, <^{\mathfrak{A}})$ we set

$$\text{field } <^{\mathfrak{A}} = \{a \in A \mid \text{for some } b \in A, a <^{\mathfrak{A}} b \text{ or } b <^{\mathfrak{A}} a\},$$[3]

then, for $(\mathbb{C}, <^{\mathbb{C}})$, field $<^{\mathbb{C}} = \mathbb{R}$ and (field $<^{\mathbb{C}}, <^{\mathbb{C}}$) is an ordering. We say that $<^{\mathfrak{A}}$ is a *partially defined ordering* on A if (field $<^{\mathfrak{A}}, <^{\mathfrak{A}}$) is a model of Φ_{ord}, i.e., if $(A, <^{\mathfrak{A}})$ satisfies

$$\Phi_{\text{pord}} \begin{cases} \forall x \, \neg x < x, \\ \forall x \, \forall y \, \forall z((x < y \wedge y < z) \rightarrow x < z), \\ \forall x \, \forall y((\exists u(x < u \vee u < x) \wedge \exists v(y < v \vee v < y)) \\ \qquad \rightarrow (x < y \vee x \equiv y \vee y < x)). \end{cases}$$

[3] Of course not to be confused with the notion of field as in 6.5.

6.5 The Theory of Fields. As symbol set let us take $S_{\mathrm{ar}} = \{+, \cdot, 0, 1\}$. An S_{ar}-structure is a *field* if it satisfies the following sentences:

$$\Phi_{\mathrm{fd}} \begin{cases} \forall x\, \forall y\, \forall z (x + y) + z \equiv x + (y + z), & \forall x\, x + 0 \equiv x, \\ \forall x\, \forall y\, \forall z (x \cdot y) \cdot z \equiv x \cdot (y \cdot z), & \forall x\, x \cdot 1 \equiv x, \\ \forall x\, \exists y\, x + y \equiv 0, & \forall x (\neg x \equiv 0 \rightarrow \exists y\, x \cdot y \equiv 1), \\ \forall x\, \forall y\, x + y \equiv y + x, & \forall x\, \forall y\, x \cdot y \equiv y \cdot x, \\ \neg 0 \equiv 1, & \\ \forall x\, \forall y\, \forall z\, x \cdot (y + z) \equiv (x \cdot y) + (x \cdot z). & \end{cases}$$

Ordered fields are $S_{\mathrm{ar}}^{<}$-structures which satisfy the following sentences:

$$\Phi_{\mathrm{ofd}} \begin{cases} \text{the sentences in } \Phi_{\mathrm{fd}} \text{ and } \Phi_{\mathrm{ord}}, \\ \forall x\, \forall y\, \forall z (x < y \rightarrow x + z < y + z), \\ \forall x\, \forall y\, \forall z ((x < y \wedge 0 < z) \rightarrow x \cdot z < y \cdot z). \end{cases}$$

6.6 Exercise. Formalize the following statements using the symbol set of 6.2:

(a) Every positive real number has a positive square root.
(b) If ρ is strictly monotone then ρ is injective.
(c) ρ is uniformly continuous on $\mathbb{R}$.
(d) For all x, if ρ is differentiable at x then ρ is continuous at x.

6.7 Exercise. Let $S_{\mathrm{eq}} = \{R\}$. Formalize:

(a) R is an equivalence relation with at least two equivalence classes.
(b) R is an equivalence relation with an equivalence class containing more than one element.

6.8 Exercise. Use 5.10 to show:

(a) If, for every $i \in I$, $\mathfrak{A}_i$ is a group then $\prod_{i \in I} \mathfrak{A}_i$ is a group.
(b) There is no set of axioms for the theory of orderings and for the theory of fields which consists of Horn sentences only.

6.9 Exercise. A set M of natural numbers is called a *spectrum* if there is a symbol set S and an S-sentence φ such that

$$M = \{n \in \mathbb{N} \mid \varphi \text{ has a model containing exactly } n \text{ elements}\}.$$

Show:

(a) Every finite subset of $\{1, 2, 3, \ldots\}$ is a spectrum.
(b) For every $m \geq 1$, the set of numbers > 0 which are divisible by m is a spectrum.

(c) The set of squares > 0 is a spectrum.
(d) The set of nonprime numbers > 0 is a spectrum.
(e) The set of prime numbers is a spectrum.

§7. Some Remarks on Formalizability

In the preceding section we had a number of examples showing how mathematical statements can be formalized by first-order formulas. However, the process of formalization is not always as simple as it was in those cases. In this section we discuss some of the typical difficulties which can arise.

7.1 Partial Functions. When we defined the notion of structure we stipulated that function symbols be interpreted by total functions, i.e., in the case of an n-ary function symbol, by a function that is defined on all n-tuples of elements of the domain. If, for example, in the field of real numbers, we regard division on $\mathbb{R}$ as a function then we do not have a structure in our sense (because the quotient is undefined if the divisor is zero). The following are possible solutions to this difficulty:

(1) The division function can be extended to a total function. For example, one can define $r/0 := 0$ for all $r \in \mathbb{R}$ and take this into consideration when formulating statements about the division function.
(2) Instead of the division function, one can consider its graph, that is, the ternary relation $\{(a, b, c) \in \mathbb{R}^3 \mid b \neq 0 \text{ and } a/b = c\}$. In VIII.1 we shall describe how statements about functions can be translated into statements about their graphs. The remarks made there for total functions can easily be modified to cover the case of partial functions.
(3) One can develop semantics for first-order languages which also include partial functions. However, this approach leads to a complicated logical system without yielding anything essentially new, as we see from (1) and (2).

7.2 Many-Sorted Structures. The structures we have hitherto considered have only one domain and in this sense consist of elements of only one *sort*. On the other hand, some important structures in mathematics contain elements of different sorts. Planes in affine spaces consist of points and lines, and vector spaces consist of vectors and scalars. Taking vector spaces as an example, we give two possibilities for treating many-sorted structures.

(1) *Many-Sorted Languages.* We regard a vector space $\mathfrak{V}$ as a "structure with two domains" (as a so-called *two-sorted structure*):

$$\mathfrak{V} = (F, V, +^F, \cdot^F, 0^F, 1^F, \oplus^V, e^V, *^{F,V}),$$

where F is the set of scalars, $(F, +^F, \cdot^F, 0^F, 1^F)$ is the field of scalars, V is the set of vectors, $(V, \oplus^V, e^V)$ is the additive group of vectors, and $*^{F,V}$ is the multiplication of scalars and vectors defined on $F \times V$.

In order to describe such two-sorted structures we introduce a two-sorted language, that is, a language built up in the same way as the languages we have used so far, but having two sorts of variables, namely $u_0, u_1, u_2, \ldots$ (for elements of the first domain, in the case above, scalars) and $v_0, v_1, v_2, \ldots$ (for elements of the second domain, in the case above, vectors).

A quantified variable always ranges over the corresponding domain. To illustrate this we formalize some of the axioms for vector spaces.

(α) Associativity of scalar addition:

$$\forall u_0 \, \forall u_1 \, \forall u_2 \, u_0 + (u_1 + u_2) \equiv (u_0 + u_1) + u_2.$$

(β) Associativity of vector addition:

$$\forall v_0 \, \forall v_1 \, \forall v_2 \, v_0 \oplus (v_1 \oplus v_2) \equiv (v_0 \oplus v_1) \oplus v_2.$$

(γ) Associativity of scalar multiplication of vectors:

$$\forall u_0 \, \forall u_1 \, \forall v_0 (u_0 \cdot u_1) * v_0 \equiv u_0 * (u_1 * v_0).$$

(2) *Sort Reduction.* It is also possible to use our one-sorted first-order languages to treat many-sorted structures, namely, by a so-called *sort reduction.* We demonstrate this method briefly for the case of vector spaces. Let $\underline{F}$ and $\underline{V}$ be two new unary relation symbols. We regard a vector space as a $\{\underline{F}, \underline{V}, +, \cdot, 0, 1, \oplus, e, *\}$-structure

$$\mathfrak{B} = (F \cup V, \underline{F}^{\mathfrak{B}}, \underline{V}^{\mathfrak{B}}, +^{\mathfrak{B}}, \cdot^{\mathfrak{B}}, 0^{\mathfrak{B}}, 1^{\mathfrak{B}}, \oplus^{\mathfrak{B}}, e^{\mathfrak{B}}, *^{\mathfrak{B}}),$$

where $\underline{F}^{\mathfrak{B}} := F$ and $\underline{V}^{\mathfrak{B}} := V$, and the functions $+^{\mathfrak{B}}, \cdot^{\mathfrak{B}}, \oplus^{\mathfrak{B}}$, and $*^{\mathfrak{B}}$ are arbitrary extensions of $+^F, \cdot^F, \oplus^V$ and $*^{F,V}$ to $(F \cup V) \times (F \cup V)$. The introduction of the "sort symbols" $\underline{F}$ and $\underline{V}$ enables us to speak of scalars and vectors. We exemplify this by reformulating the many-sorted vector axioms given above:

(α) $\forall x \, \forall y \, \forall z((\underline{F}x \wedge \underline{F}y \wedge \underline{F}z) \rightarrow (x + y) + z \equiv x + (y + z))$.
(β) $\forall x \, \forall y \, \forall z((\underline{V}x \wedge \underline{V}y \wedge \underline{V}z \rightarrow (x \oplus y) \oplus z \equiv x \oplus (y \oplus z))$.
(γ) $\forall x \, \forall y \, \forall z((\underline{F}x \wedge \underline{F}y \wedge \underline{V}z) \rightarrow (x \cdot y) * z \equiv x * (y * z))$.

Since all quantifiers are "relativized" to $\underline{F}$ or $\underline{V}$, it makes no difference how the functions $+^F, \ldots$ are extended.

7.3 Limits of Formalizability. The question of the limits of formalizability, which is ultimately the question of the expressive power of first-order languages, will be treated in detail in Chapter VI and Chapter VII, §2. Here we discuss two examples.

(1) *Torsion Groups.* A group $\mathfrak{G}$ is called a *torsion group* if every element of $\mathfrak{G}$ has finite order, i.e., if for every $a \in G$ there is an $n \geq 1$ such that $a^n = e^G$.

An ad hoc formalization of this property would be

$$\forall x(x \equiv e \vee x \circ x \equiv e \vee (x \circ x) \circ x \equiv e \vee \ldots).$$

However, in first-order logic we may not form infinitely long disjunctions. Indeed we shall later show that there is no set of first-order formulas, the models of which are precisely the torsion groups.

(2) *Peano's Axioms.* We consider the question whether there is a set of S_{ar}-sentences the models of which are the structures isomorphic to $\mathfrak{N} = (\mathbb{N}, +, \cdot, 0, 1)$. For simplicity we shall start our discussion with the structure $\mathfrak{N}_\sigma = (\mathbb{N}, \sigma, 0)$, where σ is the successor function on $\mathbb{N}$ ($\sigma(n) = n + 1$ for $n \in \mathbb{N}$). $\mathfrak{N}_\sigma$ is a $\{\underline{\sigma}, 0\}$-structure, with $\underline{\sigma}$ ("successor") a unary function symbol. The results can easily be extended to $\mathfrak{N}$, cf. Exercise 7.5.

$\mathfrak{N}_\sigma$ satisfies the so-called *Peano axiom system*:

(α) 0 is not a value of the successor function σ.
(β) σ is injective.
(γ) (the so-called *induction axiom*). For every subset X of $\mathbb{N}$: if $0 \in X$ and if $\sigma(n) \in X$ whenever $n \in X$, then $X = \mathbb{N}$.

(α) and (β) may easily be formalized in $L^{\{\underline{\sigma}, 0\}}$ by

(P1) $\forall x \neg \underline{\sigma} x \equiv 0$;
(P2) $\forall x \forall y(\underline{\sigma} x \equiv \underline{\sigma} y \rightarrow x \equiv y)$.

The induction axiom (γ) is a statement about arbitrary subsets of $\mathbb{N}$. For an "ad hoc" formalization of this axiom we would also need to quantify over variables for subsets of the domain. In such a language, (γ) could be formalized thus:

(P3) $\forall X((X0 \wedge \forall x(Xx \rightarrow X\underline{\sigma}x)) \rightarrow \forall y Xy)$.

(P3) is a so-called *second-order* formula (cf. IX.1).

The following theorem shows that (P1)–(P3) characterize the structure $\mathfrak{N}_\sigma$ up to isomorphism:

7.4 Dedekind's Theorem. *Every structure* $\mathfrak{A} = (A, \underline{\sigma}^A, 0^A)$ *which satisfies* (P1)–(P3) *is isomorphic to* $\mathfrak{N}_\sigma$.

In VI.4 we shall show that no set of first-order $\{\underline{\sigma}, 0\}$-sentences has (up to isomorphism) just $\mathfrak{N}_\sigma$ as a model. Thus the induction axiom cannot be formalized in $L^{\{\underline{\sigma}, 0\}}$.

The *proof of Dedekind's theorem* depends essentially on the fact that in structures $\mathfrak{A}$ which satisfy (P3), the following kind of inductive proofs can be given: In order to show that every element of the domain A has a certain property P, one verifies that 0^A has the property P and that if an element a has the property P then $\underline{\sigma}^A(a)$ does also.

Suppose $\mathfrak{A} = (A, \underline{\sigma}^A, 0^A)$ is a structure which satisfies (P1)–(P3). The isomorphism $\pi\colon \mathfrak{N}_\sigma \cong \mathfrak{A}$, which we need, must have the following properties:

(i) $\pi(0^{\mathbb{N}}) = 0^A$;
(ii) $\pi(\underline{\sigma}^{\mathbb{N}}(n)) = \underline{\sigma}^A(\pi(n))$ for all $n \in \mathbb{N}$,

that is,

(i)′ $\pi(0) = 0^A$;
(ii)′ $\pi(n + 1) = \underline{\sigma}^A(\pi(n))$ for all $n \in \mathbb{N}$.

We define π by induction on n, taking (i)′ and (ii)′ to be the definition. Then the compatibility conditions for an isomorphism are trivially satisfied and we need only show that π is a bijective map from $\mathbb{N}$ onto A.

Surjectivity of π: By induction in $\mathfrak{A}$ ($\mathfrak{A}$ satisfies (P3)) we prove that every element of A lies in the image of π. By (i)′, 0^A is in the image of π. Further, if a is in the image of π, say $a = \pi(n)$, then $\underline{\sigma}^A(a) = \underline{\sigma}^A(\pi(n))$; hence by (ii)′, $\underline{\sigma}^A(a) = \pi(n + 1)$, and it follows that $\underline{\sigma}^A(a)$ is also in the image of π.

Injectivity of π: By induction on n we prove

$$(*) \qquad \text{For all } m \in \mathbb{N}, \text{ if } m \neq n \text{ then } \pi(m) \neq \pi(n).$$

$n = 0$: If $m \neq 0$, say $m = k + 1$, then $\pi(m) = \pi(k + 1) = \underline{\sigma}^A(\pi(k))$, and since $\mathfrak{A}$ satisfies the axiom (P1), $\underline{\sigma}^A(\pi(k)) \neq 0^A$. Hence, by (i)′, $\pi(m) \neq \pi(0)$.

The induction step: Suppose $(*)$ has been proved for n and suppose $m \neq n + 1$. If $m = 0$, we argue as in the case $n = 0$ that $\pi(n + 1) \neq \pi(m) = 0^A$. If $m \neq 0$, say $m = k + 1$, then $k \neq n$ and so, by the induction hypothesis, $\pi(k) \neq \pi(n)$. By the injectivity of $\underline{\sigma}^A$ ($\mathfrak{A}$ satisfies (P2)) it follows that $\underline{\sigma}^A(\pi(k)) \neq \underline{\sigma}^A(\pi(n))$; hence from (ii)′ we have $\pi(k + 1) \neq \pi(n + 1)$, i.e., $\pi(m) \neq \pi(n + 1)$. □

7.5 Exercise. Let Π be the following system of second-order S_{ar}-sentences:

$$\forall x \, \neg x + 1 \equiv 0,$$

$$\forall x \, \forall y (x + 1 \equiv y + 1 \to x \equiv y),$$

$$\forall X ((X0 \wedge \forall x (Xx \to Xx + 1)) \to \forall y \, Xy),$$

$$\forall x \, x + 0 \equiv x,$$

$$\forall x \, \forall y \, x + (y + 1) \equiv (x + y) + 1,$$

$$\forall x \, x \cdot 0 \equiv 0,$$

$$\forall x \, \forall y \, x \cdot (y + 1) \equiv (x \cdot y) + x.$$

(a) If $\mathfrak{A} = (A, +^A, \cdot^A, 0^A, 1^A)$ is a model of Π and if $\underline{\sigma}^A\colon A \to A$ is given by $\underline{\sigma}^A(a) = a +^A 1^A$, then $(A, \underline{\sigma}^A, 0^A)$ satisfies the axioms (P1)–(P3).
(b) Show that $\mathfrak{N} = (\mathbb{N}, +, \cdot, 0, 1)$ is characterized by Π up to isomorphism.

§8. Substitution

In this section we define how to substitute a term t for a variable x in a formula φ at the places where x occurs free, thus obtaining a formula ψ. We wish to define the substitution in such a way that φ expresses the same about x as ψ does about t. First we give an example to illustrate our objective and to show why a certain amount of care is necessary.

Let

$$\varphi := \exists z\, z + z \equiv x.$$

In $\mathfrak{N}$ the formula φ says that x is even; more exactly,

$$(\mathfrak{N}, \beta) \models \varphi \quad \text{iff } \beta(x) \text{ is even.}$$

If we replace the variable x by y in φ we obtain the formula $\exists z\, z + z \equiv y$, which states that y is even. But if we replace the variable x by z, we obtain the formula $\exists z\, z + z \equiv z$, which no longer says that z is even; in fact, this formula is valid in $\mathfrak{N}$ regardless of the assignment for z (because $0 + 0 = 0$). The meaning is altered in this case because at the place where x occurred free, the variable z gets bound. On the other hand, we obtain a formula which expresses the same about z as φ does about x if we proceed as follows: First we introduce a new bound variable u in φ, and then in the formula $\exists u\, u + u = x$ thus obtained we replace x by z. It is immaterial which variable u (distinct from x and z) we choose. However, for certain technical purposes it is useful to make a fixed choice.

In the preceding example we replaced only one variable but in our exact definition we specify the procedure for simultaneously replacing several variables: With a given formula φ, *pairwise distinct variables* $x_0, \ldots, x_r$, and arbitrary terms $t_0, \ldots, t_r$, we associate a formula

$$\varphi \frac{t_0 \ldots t_r}{x_0 \ldots x_r},$$

which is said to be obtained from φ by *simultaneous substitution* of $t_0, \ldots, t_r$ for $x_0, \ldots, x_r$. The reader should note that the x_i need be replaced by t_i only if

$$x_i \in \text{free}(\varphi) \qquad \text{and} \qquad x_i \neq t_i.$$

In the following inductive definition this is explicitly taken into account in the quantifier step; in the other steps it follows immediately.

It will become apparent that it is convenient to first introduce a simultaneous substitution for terms. Let S be a fixed symbol set.

8.1 Definition

(a) $$x\frac{t_0\ldots t_r}{x_0\ldots x_r}:=\begin{cases}x, & \text{if } x\neq x_0,\ldots,x\neq x_r,\\ t_i, & \text{if } x=x_i,\end{cases}$$

(b) $$c\frac{t_0\ldots t_r}{x_0\ldots x_r}:=c,$$

(c) $$[ft'_0\ldots t'_{n-1}]\frac{t_0\ldots t_r}{x_0\ldots x_r}:=ft'_0\frac{t_0\ldots t_r}{x_0\ldots x_r}\ldots t'_{n-1}\frac{t_0\ldots t_r}{x_0\ldots x_r}.$$

For easier reading we use square brackets here and in what follows.

8.2 Definition

(a) $$[t'_0\equiv t'_1]\frac{t_0\ldots t_r}{x_0\ldots x_r}:=t'_0\frac{t_0\ldots t_r}{x_0\ldots x_r}\equiv t'_1\frac{t_0\ldots t_r}{x_0\ldots x_r},$$

(b) $$[Rt'_0\ldots t'_{n-1}]\frac{t_0\ldots t_r}{x_0\ldots x_r}:=Rt'_0\frac{t_0\ldots t_r}{x_0\ldots x_r}\ldots t'_{n-1}\frac{t_0\ldots t_r}{x_0\ldots x_r},$$

(c) $$[\neg\varphi]\frac{t_0\ldots t_r}{x_0\ldots x_r}:=\neg\left[\varphi\frac{t_0\ldots t_r}{x_0\ldots x_r}\right],$$

(d) $$(\varphi\vee\psi)\frac{t_0\ldots t_r}{x_0\ldots x_r}:=\varphi\frac{t_0\ldots t_r}{x_0\ldots x_r}\vee\psi\frac{t_0\ldots t_r}{x_0\ldots x_r}.$$

(e) Suppose $x_{i_0},\ldots,x_{i_{s-1}}$ $(i_0<\cdots<i_{s-1})$ are exactly the variables x_i among the $x_0,\ldots,x_r$ such that

$$x_i\in\text{free}(\exists x\varphi)\qquad\text{and}\qquad x_i\neq t_i.$$

Then set

$$[\exists x\varphi]\frac{t_0\ldots t_r}{x_0\ldots x_r}:=\exists u\left[\varphi\frac{t_{i_0}\ldots t_{i_{s-1}}u}{x_{i_0}\ldots x_{i_{s-1}}x}\right],$$

where u is the variable x if x does not occur in $t_{i_0},\ldots,t_{i_{s-1}}$; otherwise u is the first variable in the list $v_0, v_1, v_2,\ldots$ which does not occur in φ, $t_{i_0},\ldots,t_{i_{s-1}}$.

(By introducing the variable u we ensure that no variable occurring in $t_{i_0},\ldots,t_{i_{s-1}}$ falls within the scope of a quantifier. In case there is no x_i such that $x_i\in\text{free}(\exists x\varphi)$ and $x_i\neq t_i$ we have $s=0$, and from (e) we obtain

$$[\exists x\varphi]\frac{t_0\ldots t_r}{x_0\ldots x_r}=\exists x\left[\varphi\frac{x}{x}\right],$$

which is $\exists x\varphi$, as we shall see in 8.4(b).)

EXAMPLES. For binary P and f we have

(1) $$[Pv_0 fv_1v_2]\frac{v_2v_0v_1}{v_1v_2v_3} = Pv_0 fv_2v_0,$$

(2) $$[\exists v_0 Pv_0 fv_1v_2]\frac{v_4 fv_1v_1}{v_0v_2} = \exists v_0\left[Pv_0 fv_1v_2\frac{fv_1v_1v_0}{v_2\ \ v_0}\right] = \exists v_0 Pv_0 fv_1 fv_1v_1,$$

(3) $$[\exists v_0 Pv_0 fv_1v_2]\frac{v_0v_2v_4}{v_1v_2v_0} = \exists v_3\left[Pv_0 fv_1v_2\frac{v_0v_3}{v_1v_0}\right] = \exists v_3 Pv_3 fv_0v_2.$$

At the places where x_i occurred free in φ, we now find in

$$\varphi\frac{t_0\ldots t_r}{x_0\ldots x_r}$$

the term t_i. Hence, if free$(\varphi)\subset\{x_0,\ldots,x_r\}$ then we expect that

$$\varphi\frac{t_0\ldots t_r}{x_0\ldots x_r}$$

will hold for an interpretation $\mathfrak{I}=(\mathfrak{A},\beta)$ iff φ holds in $\mathfrak{A}$, provided we use the assignments $\mathfrak{I}(t_0)$ for $x_0,\ldots,\mathfrak{I}(t_r)$ for x_r. An exact formulation of this property is given in the following "substitution lemma" 8.3. Later we shall frequently refer to this lemma whereas we shall rarely return to the technical details of definition 8.2.[4] Before stating the lemma we generalize the definition of $\mathfrak{I}\frac{a}{x}$. Let $x_0,\ldots,x_r$ be pairwise distinct and suppose $\mathfrak{I}=(\mathfrak{A},\beta)$ is an interpretation, and $a_0,\ldots,a_r\in A$; then the assignment

$$\beta\frac{a_0\ldots a_r}{x_0\ldots x_r}$$

in $\mathfrak{A}$ and the interpretation

$$\mathfrak{I}\frac{a_0\ldots a_r}{x_0\ldots x_r}$$

are given by

$$\beta\frac{a_0\ldots a_r}{x_0\ldots x_r}(y):=\begin{cases}\beta(y) & \text{if } y\neq x_0,\ldots,y\neq x_r\\ a_i & \text{if } y=x_i\end{cases}$$

and

$$\mathfrak{I}\frac{a_0\ldots a_r}{x_0\ldots x_r}:=\left(\mathfrak{A},\beta\frac{a_0\ldots a_r}{x_0\ldots x_r}\right).$$

[4] As the substitution lemma the further results of this section are intuitively clear. The proofs are straightforward but lengthy and may be skipped by a reader already familiar with proofs by induction on terms and formulas.

8.3 Substitution Lemma. (a) *For every term t,*

$$\mathfrak{I}\left(t\,\frac{t_0 \ldots t_r}{x_0 \ldots x_r}\right) = \mathfrak{I}\,\frac{\mathfrak{I}(t_0) \ldots \mathfrak{I}(t_r)}{x_0 \ldots x_r}(t).$$

(b) *For every formula φ,*

$$\mathfrak{I} \models \varphi\,\frac{t_0 \ldots t_r}{x_0 \ldots x_r} \quad \textit{iff} \quad \mathfrak{I}\,\frac{\mathfrak{I}(t_0) \ldots \mathfrak{I}(t_r)}{x_0 \ldots x_r} \models \varphi.$$

PROOF. By induction on terms and formulas in accordance with the definitions 8.1 and 8.2. We treat some typical cases.

$t = x$: If $x \neq x_0, \ldots, x_r$, then, by 8.1(a)

$$x\,\frac{t_0 \ldots t_r}{x_0 \ldots x_r} = x,$$

and therefore

$$\mathfrak{I}\left(x\,\frac{t_0 \ldots t_r}{x_0 \ldots x_r}\right) = \mathfrak{I}(x) = \mathfrak{I}\,\frac{\mathfrak{I}(t_0) \ldots \mathfrak{I}(t_r)}{x_0 \ldots x_r}(x).$$

If $x = x_i$, then

$$x\,\frac{t_0 \ldots t_r}{x_0 \ldots x_r} = t_i$$

and hence

$$\mathfrak{I}\left(x\,\frac{t_0 \ldots t_r}{x_0 \ldots x_r}\right) = \mathfrak{I}(t_i) = \mathfrak{I}\,\frac{\mathfrak{I}(t_0) \ldots \mathfrak{I}(t_r)}{x_0 \ldots x_r}(x_i) = \mathfrak{I}\,\frac{\mathfrak{I}(t_0) \ldots \mathfrak{I}(t_r)}{x_0 \ldots x_r}(x).$$

$\varphi = Rt'_0 \ldots t'_{n-1}$:

$$\mathfrak{I} \models [Rt'_0 \ldots t'_{n-1}]\,\frac{t_0 \ldots t_r}{x_0 \ldots x_r} \quad \text{iff } \mathfrak{I}(R) \text{ holds for } \mathfrak{I}\left(t'_0\,\frac{t_0 \ldots t_r}{x_0 \ldots x_r}\right) \ldots$$

(by 8.2(b))

$$\text{iff } \mathfrak{I}(R) \text{ holds for } \mathfrak{I}\,\frac{\mathfrak{I}(t_0) \ldots \mathfrak{I}(t_r)}{x_0 \ldots x_r}(t'_0) \ldots$$

(by (a))

$$\text{iff } \mathfrak{I}\,\frac{\mathfrak{I}(t_0) \ldots \mathfrak{I}(t_r)}{x_0 \ldots x_r}(R) \text{ holds for}$$

$$\mathfrak{I}\,\frac{\mathfrak{I}(t_0) \ldots \mathfrak{I}(t_r)}{x_0 \ldots x_r}(t'_0) \ldots,$$

$$\text{iff } \mathfrak{I}\,\frac{\mathfrak{I}(t_0) \ldots \mathfrak{I}(t_r)}{x_0 \ldots x_r} \models Rt'_0 \ldots t'_{n-1}.$$

$\varphi = \exists x\psi$: As in 8.2(e), let $x_{i_0}, \ldots, x_{i_{s-1}}$ be exactly those variables x_i for which $x_i \in \text{free}(\exists x\psi)$ and $x_i \neq t_i$. Then, for u chosen as in 8.2(e),

$$\mathfrak{I} \models [\exists x\psi] \frac{t_0 \ldots t_r}{x_0 \ldots x_r}$$

$$\text{iff } \mathfrak{I} \models \exists u\left[\psi \frac{t_{i_0} \ldots t_{i_{s-1}} u}{x_{i_0} \ldots x_{i_{s-1}} x}\right]$$

$$\text{iff there is an } a \in A \text{ such that } \mathfrak{I}\frac{a}{u} \models \psi \frac{t_{i_0} \ldots t_{i_{s-1}} u}{x_{i_0} \ldots x_{i_{s-1}} x}$$

$$\text{iff there is an } a \in A \text{ such that } \left[\mathfrak{I}\frac{a}{u}\right] \frac{\mathfrak{I}\frac{a}{u}(t_{i_0}) \ldots \mathfrak{I}\frac{a}{u}(t_{i_{s-1}})\, \mathfrak{I}\frac{a}{u}(u)}{x_{i_0} \ldots x_{i_{s-1}} x} \models \psi$$

(by induction hypothesis)

$$\text{iff there is an } a \in A \text{ such that } \left[\mathfrak{I}\frac{a}{u}\right] \frac{\mathfrak{I}(t_{i_0}) \ldots \mathfrak{I}(t_{i_{s-1}}) a}{x_{i_0} \ldots x_{i_{s-1}} x} \models \psi$$

(by the coincidence lemma, since u does not occur in $t_{i_0}, \ldots, t_{i_{s-1}}$)

$$\text{iff there is an } a \in A \text{ such that } \mathfrak{I}\frac{\mathfrak{I}(t_{i_0}) \ldots \mathfrak{I}(t_{i_{s-1}}) a}{x_{i_0} \ldots x_{i_{s-1}} x} \models \psi$$

(by the coincidence lemma, since either $u = x$ or u does not occur in ψ)

$$\text{iff there is an } a \in A \text{ such that } \left[\mathfrak{I}\frac{\mathfrak{I}(t_{i_0}) \ldots \mathfrak{I}(t_{i_{s-1}})}{x_{i_0} \ldots x_{i_{s-1}}}\right] \frac{a}{x} \models \psi$$

(note that $x \neq x_{i_0}, \ldots, x \neq x_{i_{s-1}}$, because $x_{i_0}, \ldots, x_{i_{s-1}} \in \text{free}(\exists x\psi)$)

$$\text{iff} \quad \mathfrak{I}\frac{\mathfrak{I}(t_{i_0}) \ldots \mathfrak{I}(t_{i_{s-1}})}{x_{i_0} \ldots x_{i_{s-1}}} \models \exists x\psi$$

$$\text{iff} \quad \mathfrak{I}\frac{\mathfrak{I}(t_0) \ldots \mathfrak{I}(t_r)}{x_0 \ldots x_r} \models \exists x\psi$$

(since for $i \neq i_0, \ldots, i_{s-1}$, $x_i \notin \text{free}(\exists x\psi)$ or $x_i = t_i$). □

In the following lemma we collect several "syntactic" properties of substitution.

8.4. Lemma. (a) *For every permutation π of the numbers* $0, \ldots, r$,

$$\varphi \frac{t_0 \ldots t_r}{x_0 \ldots x_r} = \varphi \frac{t_{\pi(0)} \ldots t_{\pi(r)}}{x_{\pi(0)} \ldots x_{\pi(r)}}.$$

(b) *If* $0 \leq i \leq r$ *and* $x_i = t_i$, *then*

$$\varphi \frac{t_0 \ldots t_r}{x_0 \ldots x_r} = \varphi \frac{t_0 \ldots t_{i-1} t_{i+1} \ldots t_r}{x_0 \ldots x_{i-1} x_{i+1} \ldots x_r}.$$

In particular, $\varphi \frac{x}{x} = \varphi$.

(c) *For every variable* y,

(i) *if* $y \in \operatorname{var}\left(t \frac{t_0 \ldots t_r}{x_0 \ldots x_r}\right)$ *then* $(y \in \operatorname{var}(t_0) \cup \cdots \cup \operatorname{var}(t_r))$ *or* $(y \in \operatorname{var}(t)$ *and* $y \neq x_0, \ldots, y \neq x_r)$;

(ii) *if* $y \in \operatorname{free}\left(\varphi \frac{t_0 \ldots t_r}{x_0 \ldots x_r}\right)$ *then* $(y \in \operatorname{var}(t_0) \cup \cdots \cup \operatorname{var}(t_r))$ *or* $(y \in \operatorname{free}(\varphi)$ *and* $y \neq x_0, \ldots, y \neq x_r)$.

PROOF. By induction, using 8.1 and 8.2. We give two typical cases of (c).

$t = x$: Suppose

$$y \in \operatorname{var}\left(x \frac{t_0 \ldots t_r}{x_0 \ldots x_r}\right).$$

In case $x \neq x_0, \ldots, x \neq x_r$ we have

$$x \frac{t_0 \ldots t_r}{x_0 \ldots x_r} = x;$$

hence $y = x$ and so $(y \in \operatorname{var}(x)$ and $y \neq x_0, \ldots, y \neq x_r)$. In case $x = x_i$ we have

$$x_i \frac{t_0 \ldots t_r}{x_0 \ldots x_r} = t_i;$$

hence by assumption $y \in \operatorname{var}(t_i)$, that is, $y \in \operatorname{var}(t_0) \cup \cdots \cup \operatorname{var}(t_r)$.

$\varphi = \exists x \psi$: Let $s, i_0, \ldots, i_{s-1}$ and u be as in definition 8.2(e). Suppose

$$y \in \operatorname{free}\left([\exists x \psi] \frac{t_0 \ldots t_r}{x_0 \ldots x_r}\right) = \operatorname{free}\left(\exists u \left[\psi \frac{t_{i_0} \ldots t_{i_{s-1}} u}{x_{i_0} \ldots x_{i_{s-1}} x}\right]\right).$$

Then

$$y \neq u \qquad \text{and} \qquad y \in \operatorname{free}\left(\psi \frac{t_{i_0} \ldots t_{i_{s-1}} u}{x_{i_0} \ldots x_{i_{s-1}} x}\right);$$

thus, by induction hypothesis, $y \neq u$ and $(y \in \operatorname{var}(t_{i_0}) \cup \cdots \cup \operatorname{var}(t_{i_{s-1}}) \cup \{u\}$ or $y \in \operatorname{free}(\psi)$, $y \neq x_{i_0}, \ldots, y \neq x_{i_{s-1}}, y \neq x)$. Since for $i \neq i_0, \ldots, i_{s-1}$ we have $x_i \notin \operatorname{free}(\psi)$ or $x_i = t_i$, it follows that $y \in \operatorname{var}(t_0) \cup \cdots \cup \operatorname{var}(t_r)$ or $y \in \operatorname{free}(\psi)$, $y \neq x_0, \ldots, y \neq x_r$. □

8.5 Corollary. *Suppose* free$(\varphi) \subset \{x_0, \ldots, x_r\}$, *where we continue to assume that* $x_0, \ldots, x_r$ *are distinct. Then, for terms* $t_0, \ldots, t_r$ *such that* $\operatorname{var}(t_i) \subset \{v_0, \ldots, v_{n-1}\}$, *the formula*

$$\varphi \frac{t_0 \ldots t_r}{x_0 \ldots x_r}$$

is in L_n. *In particular,*

$$\varphi \frac{c_0 \ldots c_r}{x_0 \ldots x_r}$$

is a sentence. □

We call the number of connectives and quantifiers occurring in a formula φ the *rank* of φ, written $\operatorname{rk}(\varphi)$. More precisely:

8.6 Definition.

$$\begin{aligned} \operatorname{rk}(\varphi) &:= 0, \quad \text{if } \varphi \text{ is atomic,} \\ \operatorname{rk}(\neg\varphi) &:= \operatorname{rk}(\varphi) + 1, \\ \operatorname{rk}(\varphi \vee \psi) &:= \operatorname{rk}(\varphi) + \operatorname{rk}(\psi) + 1, \\ \operatorname{rk}(\exists x\varphi) &:= \operatorname{rk}(\varphi) + 1. \end{aligned}$$

From the definition of substitution one obtains immediately:

8.7 Lemma.

$$\operatorname{rk}\left(\varphi \frac{t_0 \ldots t_r}{x_0 \ldots x_r}\right) = \operatorname{rk}(\varphi).$$ □

The quantifier "there exists exactly one" can be conveniently formulated with the use of substitution. Let φ be a formula, x a variable, and y the first variable which is different from x and does not occur free in φ. Then we write $\exists^{=1}x\varphi$ ("there is exactly one x such that φ") for $\exists x(\varphi \wedge \forall y(\varphi \frac{y}{x} \rightarrow x \equiv y))$. It can easily be shown that for every interpretation $\mathfrak{I} = (\mathfrak{A}, \beta)$,

$$\mathfrak{I} \models \exists^{=1}x\varphi \quad \text{iff there is exactly one } a \in A \text{ such that } \mathfrak{I}\frac{a}{x} \models \varphi.$$

8.8 Exercise. For $n \geq 1$ give a similar definition of the quantifiers "there exist exactly n" and "there exist at most n".

8.9 Exercise. Let P and f be binary and set $x = v_0$, $y = v_1$, $u = v_2$, $v = v_3$, and $w = v_4$. Show, using definition 8.2 that

(a) $$\exists x\, \exists y(Pxu \wedge Pyv)\frac{uuu}{xyv} = \exists x\, \exists y(Pxu \wedge Pyu),$$

(b) $$\exists x\, \exists y(Pxu \wedge Pyv)\frac{v\ fuv}{u\ \ v} = \exists x\, \exists y(Pxv \wedge Pyfuv),$$

(c) $$\exists x\, \exists y(Pxu \wedge Pyv)\frac{u\ x\ fuv}{x\ u\ v} = \exists w\, \exists y(Pwx \wedge Pyfuv),$$

(d) $$[\forall x\, \exists y(Pxy \wedge Pxu) \vee \exists ufuu \equiv x]\frac{x\ fxy}{x\ u} = \forall v\, \exists w(Pvw \wedge Pvfxy) \vee \exists ufuu \equiv x.$$

8.10 Exercise. Show that if $x_0, \ldots, x_r \notin \mathrm{var}(t_0) \cup \cdots \cup \mathrm{var}(t_r)$ then

$$\models \varphi\frac{t_0 \ldots t_r}{x_0 \ldots x_r} \leftrightarrow \forall x_0 \ldots x_r\, (x_0 \equiv t_0 \wedge \cdots \wedge x_r \equiv t_r \to \varphi).$$

8.11 Exercise. Give a calculus in which the derivable strings are exactly those of the form

$$t\ x_0 \ldots x_r\ t_0 \ldots t_r\ t\frac{t_0 \ldots t_r}{x_0 \ldots x_r} \quad \text{or} \quad \varphi\ x_0 \ldots x_r\ t_0 \ldots t_r\ \varphi\frac{t_0 \ldots t_r}{x_0 \ldots x_r}.$$

(*Hint*: For (a) and (c) in 8.1 one can choose the following rules:

$$\frac{}{x\ x_0 \ldots x_r\ t_0 \ldots t_r x}, \quad \text{if } x \neq x_0, \ldots, x \neq x_r;$$

$$\frac{}{x\ x_0 \ldots x_r\ t_0 \ldots t_r t_i}, \quad \text{if } x = x_i;$$

$$\frac{\begin{array}{c} s_0 \quad x_0 \ldots x_r\ t_0 \ldots t_r \quad s'_0 \\ \vdots \\ s_{n-1}\ x_0 \ldots x_r\ t_0 \ldots t_r\ s'_{n-1} \end{array}}{fs_0 \ldots s_{n-1}\ x_0 \ldots x_r\ t_0 \ldots t_r\ fs'_0 \ldots s'_{n-1}}, \quad \text{if } f \in S \text{ and } f \text{ is } n\text{-ary}.$$

CHAPTER IV
A Sequent Calculus

In Chapter I we discussed the way in which a mathematician proceeds to develop a particular mathematical theory: In order to obtain an overview of the theory, he tries to find out what propositions follow from its axioms. To show that a proposition follows from the axioms he supplies a proof. Now that we have an exact definition of the notion of consequence, we are sufficiently equipped to give a more thorough discussion of the goals and methods in mathematics. If S is a symbol set and Φ is a set of S-sentences, we let $\Phi^{\models}$ be the set of S-sentences which are consequences of Φ. A mathematical proof of an S-sentence φ from the axioms in Φ shows that φ belongs to $\Phi^{\models}$. For example, consider the set Φ_{gr} of axioms for groups, where $S = S_{gr}$. The proof of I.1.1 then shows that the S_{gr}-sentence $\forall x \, \exists y \; y \circ x \equiv e$ belongs to $\Phi_{gr}^{\models}$. However, in view of the goals of the mathematician and the scope of his methods, a central question is whether *every* sentence in $\Phi^{\models}$ can be proved from the axioms in Φ. In order to answer this we must analyse the notion of proof. But even if we limit ourselves to statements which can be formulated in first-order logic, we encounter difficulties at the very outset of such an attempt. The difficulties arise from the fact that mathematicians do not have an exact, fixed notion of proof. A mathematician does not learn what a proof is from a list of permissible inferences; rather he gets acquainted with this notion by doing concrete proofs in the course of his mathematical education. Furthermore, the collection of commonly accepted methods of proof is continually being expanded by the addition of new variants. Last, but not least, the development of new theories often includes the invention of new proof techniques.

In view of this situation we shall not attempt to give an exact description of the whole spectrum of mathematical arguments. Rather we shall look at some concrete proofs and try to abstract from them certain basic constituents.

From these constituents we shall build up a precise notion of proof. It will turn out that they are sufficient to reconstruct all types of mathematical arguments. Thus we proceed as we did when we introduced the precise notion of mathematical statement, where instead of trying to give an exact description we used the first-order languages to give a clearly defined framework. In the case of first-order languages we shall merely be able to make it plausible that, in spite of their limited expressive power, these languages are in principle sufficient for the purpose of mathematics (cf. VII.2). By contrast, we can really prove that every sentence in $\Phi^{\models}$ is provable from sentences in Φ in the precise sense.

How can we single out basic constituents of mathematical deductions? If we analyse the proofs in Chapter I, for example, we see that those steps which are directly related to the meaning of the connectives, the quantifiers, and the equality symbol seem very elementary. We mention three examples. In a proof one can proceed from statements φ and ψ, which have already been obtained, to the conjunction $(\varphi \wedge \psi)$; similarly one can proceed from Pt to $\exists x\, Px$ and from Px and $x \equiv t$ to Pt. We can represent these rules schematically as follows:

$$(*) \qquad \frac{\varphi, \quad \psi}{(\varphi \wedge \psi)}, \qquad \frac{Pt}{\exists x\, Px}, \qquad \frac{Px, x \equiv t}{Pt}.$$

Written in this way, these constituents of proofs can be regarded as syntactic operations on strings of symbols. Adhering consistently to this point of view, we shall set up a list of deduction rules (in Sections 2 and 4) in this way obtaining a *calculus* $\mathfrak{S}$. We shall motivate its form in §1. In §6 (with a preview in §1) we shall give the fundamental definition for the notion of a formula φ being *formally provable* from a set Φ of formulas. This definition will be based on the notion of derivability in $\mathfrak{S}$. Formal provability is the syntactic counterpart of the semantic notion of consequence.

Throughout this chapter we fix a symbol set S.

§1. Sequent Rules

A mathematical proof proceeds from one statement to the next until it finally arrives at the assertion of the theorem in question. The individual statements depend on certain hypotheses. These can either be hypotheses of the theorem or additional hypotheses temporarily assumed in the course of the proof. For example, if one wants to prove an intermediate claim φ by contradiction one adds $\neg\varphi$ to the hypotheses; if a contradiction results then φ has been proved, and the additional assumption $\neg\varphi$ is dropped.

This observation leads us to describe a stage in a proof by listing the corresponding assumptions and the respective claim. If we call a nonempty list (sequence) of formulas a *sequent*, then we can use sequents to describe stages in a proof. For instance, the stage with assumptions $\varphi_0 \ldots \varphi_{n-1}$ and claim φ is rendered by the sequent $\varphi_0 \ldots \varphi_{n-1}\varphi$. The sequence $\varphi_0 \ldots \varphi_{n-1}$ is called the *antecedent* and φ the *succedent* of the sequent $\varphi_0 \ldots \varphi_{n-1}\varphi$. From II.4.3 it follows that the formulas which constitute a sequent are uniquely determined. In particular, the antecedent and the succedent are well-defined.

In terms of sequents, the indirect proof sketched above can be represented schematically as follows:

$$(+) \qquad \frac{\begin{array}{lll}\varphi_0 \ldots \varphi_{n-1} & \neg\varphi & \psi \\ \varphi_0 \ldots \varphi_{n-1} & \neg\varphi & \neg\psi\end{array}}{\begin{array}{lll}\varphi_0 \ldots \varphi_{n-1} & & \varphi\end{array}}.$$

Thus (+) describes the following argument: If under the assumptions $\varphi_0, \ldots, \varphi_{n-1}$ and (the additional assumption) $\neg\varphi$ one can obtain both the formula ψ and its negation $\neg\psi$ (that is, a contradiction), then from the assumptions $\varphi_0, \ldots, \varphi_{n-1}$ one can infer φ.

In the following we shall use the letters $\Gamma, \Delta, \ldots$ to denote (possibly empty) sequences of formulas. Then we can write sequents in the form $\Gamma\varphi\psi, \Delta\psi, \ldots$ and the scheme (+) as

$$(++) \qquad \frac{\begin{array}{lll}\Gamma & \neg\varphi & \psi \\ \Gamma & \neg\varphi & \neg\psi\end{array}}{\begin{array}{lll}\Gamma & & \varphi\end{array}}.$$

(As in (+), we use spaces between elements in a sequent merely for easier reading.)

According to the concepts which we have developed so far, each step in a proof leads from certain stages already attained to a new one and hence from sequents to a new sequent. Thus it seems natural to represent deduction rules such as (++) as rules of a calculus $\mathfrak{S}$, which operates on sequents (*sequent calculus*). Our conception of $\mathfrak{S}$ is based upon [16]. For comparison the reader can find calculi of a different nature in [25].

Before listing the rules of $\mathfrak{S}$ in the next section, some further remarks will be helpful.

If, in the calculus $\mathfrak{S}$, there is a derivation of the sequent $\Gamma\varphi$, then we write $\vdash \Gamma\varphi$ and say that $\Gamma\varphi$ is derivable.

1.1 Definition. A formula φ is *formally provable* or *derivable* from a set Φ of formulas (written: $\Phi \vdash \varphi$) if and only if there are finitely many formulas $\varphi_0, \ldots, \varphi_{n-1}$ in Φ such that $\vdash \varphi_0 \ldots \varphi_{n-1}\varphi$.

A sequent $\Gamma\varphi$ is called *correct* if $\Gamma \models \varphi$, more precisely, if $\{\psi \mid \psi$ is a member of $\Gamma\} \models \varphi$. Since the rules of $\mathfrak{S}$ are modelled after usual mathematical inferences, it will turn out that they are *correct*, i.e., when applied to correct sequents they yield a correct sequent. As a result, every formula which is derivable from Φ also follows from Φ. We convince ourselves of the correctness of each rule as soon as we introduce it.

§2. Structural Rules and Connective Rules

We divide the rules of the sequent calculus $\mathfrak{S}$ into the following categories: *structural rules* (2.1, 2.2), *connective rules* (2.3, 2.4, 2.5, 2.6), *quantifier rules* (4.1, 4.2), and *equality rules* (4.3, 4.4). We start with the two structural rules.

2.1 Antecedent Rule (Ant).

$$\frac{\Gamma \quad \varphi}{\Gamma' \quad \varphi}, \quad \textit{if every member of } \Gamma \textit{ is also a member of } \Gamma' \textit{ (briefly: if } \Gamma \subset \Gamma'\textit{)}.$$

Note that a formula which occurs more than once in Γ need only occur once in Γ'.

2.2 Assumption Rule (Ass).

$$\frac{}{\Gamma \quad \varphi}, \textit{ if } \varphi \textit{ is a member of } \Gamma.$$

CORRECTNESS. (Ant): If a sequent $\Gamma\varphi$ is correct and $\Gamma \subset \Gamma'$, then since $\Gamma \models \varphi$, also $\Gamma' \models \varphi$. (Ass) is correct since $\Phi \models \varphi$ always holds for $\varphi \in \Phi$. □

(Ass) reflects the trivial fact that one can conclude φ from a set of assumptions which includes φ. (Ant) expresses the fact that one can re-order or add to assumptions.

The first negation rule incorporates the commonly used method of *proof by cases*. In order to conclude φ from Γ one first considers the case where a condition ψ holds and then treats the case where $\neg\psi$ holds. That is, one first has ψ and then $\neg\psi$ as an additional assumption. We can translate this argument into a rule for sequents as follows:

2.3 Proof by Cases Rule (PC).

$$\begin{array}{l} \Gamma \quad \psi \quad \varphi \\ \Gamma \quad \neg\psi \quad \varphi \\ \hline \Gamma \qquad\quad \varphi \end{array}.$$

CORRECTNESS. Suppose $\Gamma\,\psi \models \varphi$ and $\Gamma\,\neg\psi \models \varphi$ hold. We must show that $\Gamma \models \varphi$. Let $\mathfrak{I}$ be any interpretation such that $\mathfrak{I} \models \Gamma$, i.e., $\mathfrak{I} \models \chi$ for every member χ of Γ. Either $\mathfrak{I} \models \psi$ or $\mathfrak{I} \models \neg\psi$. If $\mathfrak{I} \models \psi$ then since $\Gamma\,\psi \models \varphi$ it follows that $\mathfrak{I} \models \varphi$. If $\mathfrak{I} \models \neg\psi$ one obtains the same result because $\Gamma\,\neg\psi \models \varphi$. □

As the second rule concerning negation we take the schema $(++)$ given in §1:

2.4 Contradiction Rule (Ctr).

$$\begin{array}{lll} \Gamma & \neg\varphi & \psi \\ \Gamma & \neg\varphi & \neg\psi \\ \hline \Gamma & & \varphi \end{array}.$$

CORRECTNESS. Let $\Gamma\ \neg\varphi \models \psi$ and $\Gamma\ \neg\varphi \models \neg\psi$. Then there is no interpretation satisfying $\Gamma\ \neg\varphi$; hence any interpretation satisfying Γ must satisfy φ, i.e., $\Gamma\ \varphi$ is correct. □

2.5 $\vee$-Rule for the Antecedent ($\vee$A)

$$\begin{array}{lll} \Gamma & \varphi & \chi \\ \Gamma & \psi & \chi \\ \hline \Gamma & (\varphi \vee \psi) & \chi \end{array}.$$

The proof that this rule is correct is similar to that for (PC).

2.6 $\vee$-Rules for the Succedent ($\vee$S)

(a) $\dfrac{\Gamma\ \varphi}{\Gamma\ (\varphi \vee \psi)}$; (b) $\dfrac{\Gamma\ \varphi}{\Gamma\ (\psi \vee \varphi)}$.

CORRECTNESS. Suppose $\Gamma \models \varphi$ and let $\mathfrak{I} \models \Gamma$. Then $\mathfrak{I} \models \varphi$ and hence both $\mathfrak{I} \models (\varphi \vee \psi)$ and $\mathfrak{I} \models (\psi \vee \varphi)$. □

2.7 Exercise. Decide whether the following rules are correct:

(a) $\begin{array}{ll} \Gamma\ \varphi_1 & \psi_1 \\ \Gamma\ \varphi_2 & \psi_2 \\ \hline \Gamma\ (\varphi_1 \vee \varphi_2) & (\psi_1 \vee \psi_2) \end{array}$; (b) $\begin{array}{ll} \Gamma\ \varphi_1 & \psi_1 \\ \Gamma\ \varphi_2 & \psi_2 \\ \hline \Gamma\ (\varphi_1 \vee \varphi_2) & (\psi_1 \wedge \psi_2) \end{array}$.

§3. Derivable Connective Rules

Using the rules of $\mathfrak{S}$ which we have formulated so far, we derive a number of sequents and introduce the notion of a *derivable rule*. In our first example we show that all sequents of the form $(\varphi \vee \neg\varphi)$ are derivable. Our notation is similar to that used for derivations in previous calculi (cf. Chapter II, §3).

	1.	φ	φ	(Ass)
	2.	φ	$(\varphi \vee \neg\varphi)$	($\vee$S) applied to 1
(∗)	3.	$\neg\varphi$	$\neg\varphi$	(Ass)
	4.	$\neg\varphi$	$(\varphi \vee \neg\varphi)$	($\vee$S) applied to 3
	5.		$(\varphi \vee \neg\varphi)$	(PC) applied to 2 and 4.

We consider the rule (TND) ("Tertium non datur" or "Law of the excluded middle")

$$\overline{(\varphi \vee \neg\varphi)},$$

which is not a rule of $\mathfrak{S}$. If we add (TND) to $\mathfrak{S}$ we do not enlarge the set of derivable sequents. For if we are given a derivation of a sequent which uses rules of $\mathfrak{S}$ together with (TND), we can insert lines 1–4 of (*) directly before every sequent $(\varphi \vee \neg\varphi)$, which originally was introduced by (TND). In this way we obtain a derivation in $\mathfrak{S}$.

Rules for sequents, whose use in a derivation can be eliminated by a derivation schema like (*), and which therefore do not enlarge the set of derivable formulas, will be called *derivable rules*. Thus (TND) is a derivable rule. The use of such derivable rules contributes to the transparency of derivations in the sequent calculus. In the remainder of this section we give some useful examples, also including derivable rules with premises.

3.1 Second Contradiction Rule (Ctr′).

$$\begin{array}{l} \Gamma\ \psi \\ \Gamma\ \neg\psi \\ \hline \Gamma\ \varphi \end{array}$$

Justification. (The justification shows that the rule is derivable. In this case we have to show how one can use rules of $\mathfrak{S}$ to obtain the sequent $\Gamma\ \varphi$ from (the "premises") $\Gamma\ \psi$ and $\Gamma\ \neg\psi$.)

1. $\Gamma \quad \psi$ premise
2. $\Gamma \quad \neg\psi$ premise
3. $\Gamma\ \neg\varphi\ \psi$ (Ant) applied to 1
4. $\Gamma\ \neg\varphi\ \neg\psi$ (Ant) applied to 2
5. $\Gamma \quad \varphi$ (Ctr) applied to 3 and 4.

3.2 Chain Rule (Ch).

$$\begin{array}{l} \Gamma\ \quad \varphi \\ \Gamma\ \varphi\ \psi \\ \hline \Gamma\ \quad \psi \end{array}$$

Justification.

1. $\Gamma \quad \varphi$ premise
2. $\Gamma\ \varphi \quad \psi$ premise
3. $\Gamma\ \neg\varphi\ \varphi$ (Ant) applied to 1

4. $\Gamma\ \neg\varphi\ \neg\varphi$ (Ass)
5. $\Gamma\ \neg\varphi\ \psi$ (Ctr′) applied to 3 and 4
6. $\Gamma\ \ \psi$ (PC) applied to 2 and 5.

3.3 Contraposition Rules (Cp).

(a) $\dfrac{\Gamma\ \ \varphi\ \ \psi}{\Gamma\ \neg\psi\ \neg\varphi}$; (b) $\dfrac{\Gamma\ \neg\varphi\ \neg\psi}{\Gamma\ \ \psi\ \ \varphi}$; (c) $\dfrac{\Gamma\ \neg\varphi\ \psi}{\Gamma\ \neg\psi\ \varphi}$; (d) $\dfrac{\Gamma\ \varphi\ \neg\psi}{\Gamma\ \psi\ \neg\varphi}$.

JUSTIFICATION OF (*a*)

1. $\Gamma\ \varphi\ \ \psi$ premise
2. $\Gamma\ \neg\psi\ \varphi\ \ \psi$ (Ant) applied to 1
3. $\Gamma\ \neg\psi\ \varphi\ \ \neg\psi$ (Ass)
4. $\Gamma\ \neg\psi\ \varphi\ \ \neg\varphi$ (Ctr′) applied to 2 and 3
5. $\Gamma\ \neg\psi\ \neg\varphi\ \neg\varphi$ (Ass)
6. $\Gamma\ \neg\psi\ \ \neg\varphi$ (PC) applied to 4 and 5.

3.4

$$\begin{array}{l}\Gamma\ (\varphi \vee \psi)\\ \Gamma\ \neg\varphi\\ \hline \Gamma\ \psi\end{array}.$$

JUSTIFICATION

1. $\Gamma\ \ (\varphi \vee \psi)$ premise
2. $\Gamma\ \ \neg\varphi$ premise
3. $\Gamma\ \psi\ \ \psi$ (Ass)
4. $\Gamma\ \varphi\ \ \neg\varphi$ (Ant) applied to 2
5. $\Gamma\ \varphi\ \ \varphi$ (Ass)
6. $\Gamma\ \varphi\ \ \psi$ (Ctr′) applied to 5 and 4
7. $\Gamma\ (\varphi \vee \psi)\ \psi$ ($\vee$A) applied to 6 and 3
8. $\Gamma\ \ \psi$ (Ch) applied to 1 and 7.

3.5 "Modus ponens".

$$\begin{array}{l}\Gamma\ (\varphi \to \psi)\\ \Gamma\ \varphi\\ \hline \Gamma\ \psi\end{array}, \quad \text{that is,} \quad \begin{array}{l}\Gamma\ (\neg\varphi \vee \psi)\\ \Gamma\ \varphi\\ \hline \Gamma\ \psi\end{array}.$$

The following justification of 3.5 is analogous to the one given for 3.4.

JUSTIFICATION

1.	Γ	$(\neg\varphi \vee \psi)$	premise
2.	Γ	φ	premise
3.	$\Gamma\ \neg\varphi$	φ	(Ant) applied to 2
4.	$\Gamma\ \neg\varphi$	$\neg\varphi$	(Ass)
5.	$\Gamma\ \neg\varphi$	ψ	(Ctr′) applied to 3 and 4
6.	$\Gamma\ \psi$	ψ	(Ass)
7.	$\Gamma\ (\neg\varphi \vee \psi)$	ψ	$(\vee\text{A})$ applied to 5 and 6
8.	Γ	ψ	(Ch) applied to 1 and 7.

Using the preceding rules we obtain:

3.6 Lemma. *The following sequents are derivable*:

(a1) $\varphi\ (\varphi \vee \psi)$; (a2) $\psi\ (\varphi \vee \psi)$;
(b) $(\varphi \vee \psi)\ \neg\varphi\ \psi$; (c) $(\neg\varphi \vee \psi)\ \varphi\ \psi$.

PROOF. For (a1):

1. $\varphi\ \varphi$ (Ass)
2. $\varphi\ (\varphi \vee \psi)$ $(\vee\text{S})$ applied to 1.

(a2), (b), and (c) can be proved similarly by using $(\vee\text{S})$, 3.4, or 3.5, respectively. □

3.7 Exercise. Show that the following rules are derivable.

(a1) $\dfrac{\Gamma\ \varphi}{\Gamma\ \neg\neg\varphi}$; (a2) $\dfrac{\Gamma\ \neg\neg\varphi}{\Gamma\ \varphi}$;

(b) $\dfrac{\begin{matrix}\Gamma\ \varphi\\ \Gamma\ \psi\end{matrix}}{\Gamma\ (\varphi \wedge \psi)}$, that is, $\dfrac{\begin{matrix}\Gamma\ \varphi\\ \Gamma\ \psi\end{matrix}}{\Gamma\ \neg(\neg\varphi \vee \neg\psi)}$;

(c1) $\dfrac{\Gamma\ (\varphi \wedge \psi)}{\Gamma\ \varphi}$; (c2) $\dfrac{\Gamma\ (\varphi \wedge \psi)}{\Gamma\ \psi}$;

(d) $\dfrac{\Gamma\ \varphi\ \psi}{\Gamma\quad (\varphi \rightarrow \psi)}$.

§4. Quantifier and Equality Rules

Now we give two sequent rules of $\mathfrak{S}$ which involve the existential quantifier. The first is a generalization of a scheme already mentioned in the introduction to this chapter.

4.1 Rule for $\exists$-Introduction in the Succedent ($\exists$S).

$$\frac{\Gamma \ \varphi \frac{t}{x}}{\Gamma \ \exists x \varphi}.$$

($\exists$S) says, in essence, that we can conclude $\exists x \varphi$ from Γ if we have already obtained the "witness" t for this existence claim.

Correctness. Suppose $\Gamma \models \varphi \frac{t}{x}$. Let $\mathfrak{I}$ be an interpretation such that $\mathfrak{I} \models \Gamma$. By assumption $\mathfrak{I} \models \varphi \frac{t}{x}$ holds. Then by the substitution lemma, $\mathfrak{I} \frac{\mathfrak{I}(t)}{x} \models \varphi$ and hence $\mathfrak{I} \models \exists x \varphi$ also. □

The second $\exists$-rule is more complicated, but it incorporates a method of argument that is used frequently. The aim is to prove a claim ψ from assumptions $\varphi_0, \ldots, \varphi_{n-1}, \exists x \varphi$. (On our formal level to achieve the corresponding aim requires a derivation of the sequent

$$(*) \qquad \varphi_0 \ldots \varphi_{n-1} \exists x \varphi \ \psi$$

in the sequent calculus.)

According to the hypothesis $\exists x \varphi$, one assumes one has an example—denoted by a new variable y—which "satisfies φ" and uses it to prove ψ. (In the sequent calculus this corresponds to a derivation of

$$(**) \qquad \varphi_0 \ldots \varphi_{n-1} \varphi \frac{y}{x} \ \psi,$$

where y is not free in $(*)$.) Then one regards ψ as having been proved from $\varphi_0, \ldots, \varphi_{n-1}, \exists x \varphi$[1]. We can reproduce this argument in the sequent calculus by a rule which allows us to proceed from $(**)$ to $(*)$:

4.2 Rule for $\exists$-Introduction in the Antecedent ($\exists$A).

$$\frac{\Gamma \ \varphi \frac{y}{x} \ \psi}{\Gamma \ \exists x \varphi \ \psi}, \quad \text{if } y \text{ is not free in } \Gamma \ \exists x \varphi \ \psi.$$

[1] Cf. the proof of I.1.1. with the use of y in line (1).

CORRECTNESS. Suppose $\Gamma\, \varphi \frac{y}{x} \models \psi$, y is not free in $\Gamma\, \exists x \varphi\, \psi$, and the interpretation $\mathfrak{I} = (\mathfrak{A}, \beta)$ is a model of $\Gamma\, \exists x \varphi$. We must show that $\mathfrak{I} \models \psi$. First, there is an $a \in A$ such that $\mathfrak{I} \frac{a}{x} \models \varphi$. Using the coincidence lemma we can conclude $\left(\mathfrak{I} \frac{a}{y}\right) \frac{a}{x} \models \varphi$. (For $x = y$ this is clear; for $x \neq y$ note that $y \notin \text{free}(\varphi)$ since otherwise $y \in \text{free}(\exists x \varphi)$ contrary to assumption.) Because $\mathfrak{I} \frac{a}{y}(y) = a$ we have

$$\left(\mathfrak{I} \frac{a}{y}\right) \frac{\mathfrak{I} \frac{a}{y}(y)}{x} \models \varphi$$

and hence by the substitution lemma $\mathfrak{I} \frac{a}{y} \models \varphi \frac{y}{x}$. From $\mathfrak{I} \models \Gamma$ and $y \notin \text{free}(\Gamma)$ we get $\mathfrak{I} \frac{a}{y} \models \Gamma$, again by the coincidence lemma; since $\Gamma \varphi \frac{y}{x} \models \psi$ we obtain $\mathfrak{I} \frac{a}{y} \models \psi$ and therefore $\mathfrak{I} \models \psi$ because $y \notin \text{free}(\psi)$. $\square$

The condition on y in ($\exists$A) is essential. For example, the sequent $[x \equiv fy] \frac{y}{x}\; y \equiv fy$ is correct; however, the sequent $\exists x\, x \equiv fy \quad y \equiv fy$, which we could obtain by applying ($\exists$A) while ignoring this extra condition, is no longer correct. This can be verified, say, by an interpretation with domain $\mathbb{N}$, which interprets f as the successor function $n \mapsto n + 1$ and y as 0.

From a formula $\varphi \frac{t}{x}$ it is not in general possible to recover either φ or t. For instance, the formula Rfy can be written as $Rx \frac{fy}{x}$ or as $Rfx \frac{y}{x}$. Therefore, in applications of the rules ($\exists$S) and ($\exists$A), we shall explicitly mention φ and t or φ and y if they are not clear from the notation.

The last two rules of $\mathfrak{S}$ arise from two basic properties of the equality relation.

4.3 Reflexivity Rule for Equality ($\equiv$).

$$\overline{t \equiv t}\,.$$

4.4 Substitution Rule for Equality (Sub).

$$\frac{\Gamma \qquad \varphi\frac{t}{x}}{\Gamma\ t \equiv t'\ \varphi\frac{t'}{x}}$$

CORRECTNESS. ($\equiv$): trivial. (Sub): Suppose $\Gamma \models \varphi\frac{t}{x}$ and suppose $\mathfrak{I}$ satisfies Γ and $t \equiv t'$. Then $\mathfrak{I} \models \varphi\frac{t}{x}$ and hence, by the substitution lemma, $\mathfrak{I}\frac{\mathfrak{I}(t)}{x} \models \varphi$; therefore since $\mathfrak{I}(t) = \mathfrak{I}(t')$ we have $\mathfrak{I}\frac{\mathfrak{I}(t')}{x} \models \varphi$. A further application of the substitution lemma yields finally that $\mathfrak{I} \models \varphi\frac{t'}{x}$. □

4.5 Exercise. Decide whether the following rules are correct:

(a) $\frac{\varphi \quad \psi}{\exists x\varphi\ \exists x\psi}$; (b) $\frac{\Gamma\ \varphi \quad \psi}{\Gamma\ \exists x\varphi\ \exists x\psi}$;

(c) $\frac{\Gamma\ \varphi\frac{fy}{x}}{\Gamma\ \forall x\varphi}$, if f is unary, and f and y do not occur in $\Gamma\ \forall x\varphi$.

§5. Further Derivable Rules and Sequents

Since $\varphi\frac{x}{x} = \varphi$, we obtain from 4.1 and 4.2 (for $t = x$ and $y = x$)

5.1.

(a) $\frac{\Gamma\ \varphi}{\Gamma\ \exists x\varphi}$,

(b) $\frac{\Gamma\ \varphi \quad \psi}{\Gamma\ \exists x\varphi\ \psi}$, if x is not free in $\Gamma\ \psi$.

A corresponding special case of (Sub) is

5.2.

$$\frac{\Gamma \qquad \varphi}{\Gamma\ x \equiv t\ \varphi\frac{t}{x}}.$$

We conclude with some derivable sequents dealing with the *symmetry* and the *transitivity* of the equality relation and its *compatibility* with functions and relations.

5.3.

$$\vdash t_0 \equiv t_1 \quad t_1 \equiv t_0.$$

5.4.

$$\vdash t_0 \equiv t_1 \quad t_1 \equiv t_2 \quad t_0 \equiv t_2.$$

5.5. For every $R \in S$, R n-ary,

$$\vdash Rt_0 \ldots t_{n-1} \quad t_0 \equiv t'_0 \quad t_1 \equiv t'_1 \quad \ldots \quad t_{n-1} \equiv t'_{n-1} \quad Rt'_0 \ldots t'_{n-1}.$$

5.6. For every $f \in S$, f n-ary,

$$\vdash t_0 \equiv t'_0 \quad t_1 \equiv t'_1 \quad \ldots \quad t_{n-1} \equiv t'_{n-1} \quad ft_0 \ldots t_{n-1} \equiv ft'_0 \ldots t'_{n-1}.$$

JUSTIFICATION OF 5.3 THROUGH 5.6

5.3: Let x be a variable not occurring in t_0 or t_1.

1. $\qquad t_0 \equiv t_0 \quad (\equiv)$

2. $t_0 \equiv t_1 \quad t_1 \equiv t_0$ (Sub) applied to 1 using $t_0 \equiv t_0 = [x \equiv t_0]\frac{t_0}{x}$.

5.4: Suppose x is a variable not occurring in t_0, t_1, or t_2.

1. $t_0 \equiv t_1 \qquad t_0 \equiv t_1$ (Ass)

2. $t_0 \equiv t_1 \quad t_1 \equiv t_2 \quad t_0 \equiv t_2$ (Sub) applied to 1 using $t_0 \equiv t_1 = [t_0 \equiv x]\frac{t_1}{x}$.

5.5: For simplicity we assume that $n = 2$. Let x be a variable which does not occur in t_0, t_1, t'_0, or t'_1.

1. $Rt_0t_1 \qquad Rt_0t_1$ (Ass)

2. $Rt_0t_1 \quad t_0 \equiv t'_0 \qquad Rt'_0t_1$ (Sub) applied to 1 using

$$Rt_0t_1 = [Rxt_1]\frac{t_0}{x}$$

3. $Rt_0t_1 \quad t_0 \equiv t'_0 \quad t_1 \equiv t'_1 \quad Rt'_0t'_1$ (Sub) applied to 2. using

$$Rt'_0t_1 = [Rt'_0x]\frac{t_1}{x}.$$

5.6 can be treated similarly. □

5.7 Exercise. Show that the following rules are derivable:

(a1) $\dfrac{\Gamma \ \forall x\varphi}{\Gamma \ \varphi\frac{t}{x}}$, that is, $\dfrac{\Gamma \neg \exists x \neg \varphi}{\Gamma \ \varphi\frac{t}{x}}$; (a2) $\dfrac{\Gamma \ \forall x\varphi}{\Gamma \quad \varphi}$.

(b1) $\dfrac{\Gamma \ \varphi\frac{t}{x} \ \psi}{\Gamma \ \forall x\varphi \ \psi}$; (b2) $\dfrac{\Gamma \ \varphi\frac{y}{x}}{\Gamma \ \forall x\varphi}$, if y is not free in $\Gamma \ \forall x\varphi$;

(b3) $\dfrac{\Gamma \quad \varphi \ \psi}{\Gamma \ \forall x\varphi \ \psi}$; (b4) $\dfrac{\Gamma \quad \varphi}{\Gamma \ \forall x\varphi}$, if x is not free in Γ.

§6. Summary and Example

For the reader's convenience we list all the rules of $\mathfrak{S}$ together:

(Ass) $\dfrac{}{\Gamma \ \varphi}$, if φ is in Γ

(Ant) $\dfrac{\Gamma \ \varphi}{\Gamma' \ \varphi}$, if $\Gamma \subset \Gamma'$

(PC) $\dfrac{\begin{array}{ll}\Gamma & \psi \ \varphi \\ \Gamma & \neg\psi \ \varphi\end{array}}{\Gamma \quad \varphi}$

(Ctr) $\dfrac{\begin{array}{ll}\Gamma & \neg\varphi \quad \psi \\ \Gamma & \neg\varphi \ \neg\psi\end{array}}{\Gamma \quad \varphi}$

($\vee$A) $\dfrac{\begin{array}{ll}\Gamma & \varphi \quad \chi \\ \Gamma & \psi \quad \chi\end{array}}{\Gamma \ (\varphi \vee \psi) \ \chi}$

($\vee$S) $\dfrac{\Gamma \quad \varphi}{\Gamma \ (\varphi \vee \psi)}, \dfrac{\Gamma \quad \varphi}{\Gamma \ (\psi \vee \varphi)}$

($\exists$A) $\dfrac{\Gamma \ \varphi\frac{y}{x} \ \psi}{\Gamma \ \exists x\varphi \ \psi}$, if y is not free in $\Gamma \ \exists x\varphi \ \psi$

($\exists$S) $\dfrac{\Gamma \ \varphi\frac{t}{x}}{\Gamma \ \exists x\varphi}$

($\equiv$) $\dfrac{}{t \equiv t}$

(Sub) $\dfrac{\Gamma \qquad \varphi\frac{t}{x}}{\Gamma \ t \equiv t' \ \varphi\frac{t'}{x}}$.

In 1.1 we defined a formula φ to be *derivable* (*formally provable*) from Φ (written: $\Phi \vdash \varphi$) if there are formulas $\varphi_0, \ldots, \varphi_{n-1}$ in Φ such that $\vdash \varphi_0 \ldots \varphi_{n-1}\varphi$. From this definition we immediately obtain:

6.1 Lemma. *For all Φ and φ, $\Phi \vdash \varphi$ if and only if there is a* finite *subset Φ_0 of Φ such that $\Phi_0 \vdash \varphi$.* □

We have already more or less proved the correctness of $\mathfrak{S}$:

6.2 Theorem on the Correctness of $\mathfrak{S}$. *For all Φ and φ, if $\Phi \vdash \varphi$, then $\Phi \models \varphi$.*

Proof. Suppose $\Phi \vdash \varphi$. Then for a suitable Γ from Φ (that is, a Γ whose members are formulas from Φ) we have $\vdash \Gamma\ \varphi$. As we showed, every rule without premises yields only correct sequents, and the other rules of $\mathfrak{S}$ always lead from correct sequents to correct sequents. Thus, by induction over $\mathfrak{S}$, we see that every derivable sequent is correct, hence also $\Gamma\ \varphi$. Therefore $\Gamma \models \varphi$ and so $\Phi \models \varphi$. □

We shall prove the converse of 6.2, namely, "if $\Phi \models \varphi$ then $\Phi \vdash \varphi$", in the next chapter. In particular, it will follow that if φ is *mathematically* provable from Φ, and hence $\Phi \models \varphi$, then φ is also *formally* provable from Φ. However, because of the elementary character of the rules for sequents, a formal proof is in general considerably more complicated than the corresponding mathematical proof. As an example we give here a formal proof of the theorem $\forall x\, \exists y\, y \circ x \equiv e$ (existence of a left inverse) from the group axioms $\varphi_0 = \forall x\, \forall y\, \forall z\, (x \circ y) \circ z \equiv x \circ (y \circ z)$, $\varphi_1 = \forall x\, x \circ e \equiv x$, and $\varphi_2 = \forall x\, \exists y\, x \circ y \equiv e$. For simplicity we shall write xy instead of $x \circ y$. The reader should compare the formal proof below with the mathematical proof of the same theorem in I.1.1. The chain of equations given there corresponds to the underlined formulas in the derivation. For easier reading we use the derivable rules:

$$(\text{Sym})\ \frac{\Gamma\ t_0 \equiv t_1}{\Gamma\ t_1 \equiv t_0} \qquad \text{and} \qquad (\text{Trans})\ \frac{\begin{array}{l}\Gamma\ t_0 \equiv t_1\\ \Gamma\ t_1 \equiv t_2\end{array}}{\Gamma\ t_0 \equiv t_2}.$$

The reader can easily justify them by means of 5.3, 5.4, and (Ch).

1. $\varphi_0\ \varphi_1\ \varphi_2$	$\forall x\, xe \equiv x$	(Ass)
2. $\varphi_0\ \varphi_1\ \varphi_2$	$(yx)e \equiv yx$	5.7(a1) applied to 1 setting $t = yx$
3. $\varphi_0\ \varphi_1\ \varphi_2$	$\underline{yx \equiv (yx)e}$	(Sym) applied to 2
4. $\varphi_0\ \varphi_1\ \varphi_2\ e \equiv yz$	$yx \equiv (yx)(yz)$	(Sub) applied to 3
5. $\varphi_0\ \varphi_1\ \varphi_2\ yz \equiv e$	$e \equiv yz$	5.3 and (Ant)
6. $\varphi_0\ \varphi_1\ \varphi_2\ yz \equiv e$	$\underline{yx \equiv (yx)(yz)}$	(Ant) and (Ch) applied to 5 and 4
7. $\varphi_0\ \varphi_1\ \varphi_2\ yz \equiv e$	$\forall x\, \forall y\, \forall z\, (xy)z \equiv x(yz)$	(Ass)
8. $\varphi_0\ \varphi_1\ \varphi_2\ yz \equiv e$	$\forall u\, \forall z\, (yu)z \equiv y(uz)$	5.7 (a1) applied to 7 setting $t = y$

9.	$\varphi_0\ \varphi_1\ \varphi_2\ yz \equiv e$	$\forall z\,(yx)z \equiv y(xz)$	5.7(a1) applied to 8 setting $t = x$
10.	$\varphi_0\ \varphi_1\ \varphi_2\ yz \equiv e$	$(yx)(yz) \equiv y(x(yz))$	5.7(a1) applied to 9 setting $t = yz$
11.	$\varphi_0\ \varphi_1\ \varphi_2\ yz \equiv e$	$\underline{yx \equiv y(x(yz))}$	(Trans) applied to 6 and 10
12.	$\varphi_0\ \varphi_1\ \varphi_2\ yz \equiv e\ \ x(yz) \equiv (xy)z$	$yx \equiv y((xy)z)$	(Sub) applied to 11
13.	$\varphi_0\ \varphi_1\ \varphi_2\ yz \equiv e$	$(xy)z \equiv x(yz)$	5.7(a2) applied three times to 7
14.	$\varphi_0\ \varphi_1\ \varphi_2\ yz \equiv e$	$x(yz) \equiv (xy)z$	(Sym) applied to 13
15.	$\varphi_0\ \varphi_1\ \varphi_2\ yz \equiv e$	$\underline{yx \equiv y((xy)z)}$	(Ch) applied to 14 and 12
16.	$\varphi_0\ \varphi_1\ \varphi_2\ yz \equiv e\ \ xy \equiv e$	$\underline{yx \equiv y(ez)}$	(Sub) applied to 15
17.	$\varphi_0\ \varphi_1\ \varphi_2\ yz \equiv e\ \ xy \equiv e$	$(ye)z \equiv y(ez)$	5.7(a1) applied three times to φ_0; like steps 7–10
18.	$\varphi_0\ \varphi_1\ \varphi_2\ yz \equiv e\ \ xy \equiv e$	$y(ez) \equiv (ye)z$	(Sym) applied to 17
19.	$\varphi_0\ \varphi_1\ \varphi_2\ yz \equiv e\ \ xy \equiv e$	$\underline{yx \equiv (ye)z}$	(Trans) applied to 16 and 18
20.	$\varphi_0\ \varphi_1\ \varphi_2\ yz \equiv e\ \ xy \equiv e\ ye \equiv y$	$yx \equiv yz$	(Sub) applied to 19
21.	$\varphi_0\ \varphi_1\ \varphi_2\ yz \equiv e\ \ xy \equiv e$	$ye \equiv y$	5.7(a1) applied to 1 setting $t = y$, and (Ant)
22.	$\varphi_0\ \varphi_1\ \varphi_2\ yz = e\ \ xy \equiv e$	$\underline{yx \equiv yz}$	(Ch) applied to 21 and 20
23.	$\varphi_0\ \varphi_1\ \varphi_2\ xy \equiv e\ \ yz \equiv e$	$\underline{yx \equiv e}$	(Sub) and (Ant) applied to 22
24.	$\varphi_0\ \varphi_1\ \varphi_2\ xy \equiv e\ \ yz \equiv e$	$\exists y\ yx \equiv e$	($\exists$S) applied to 23
25.	$\varphi_0\ \varphi_1\ \varphi_2\ xy \equiv e\ \exists z\ yz \equiv e$	$\exists y\ yx \equiv e$	($\exists$A) applied to 24

26. $\varphi_0\ \varphi_1\ \varphi_2\ xy \equiv e\ \forall y\, \exists z\, yz \equiv e$	$\exists y\, yx \equiv e$	5.7(b3) applied to 25
27. $xy \equiv e$	$xy \equiv e$	(Ass)
28. $xy \equiv e$	$\exists z\, xz \equiv e$	($\exists$S) applied to 27
29. $\exists y\, xy \equiv e$	$\exists z\, xz \equiv e$	($\exists$A) applied to 28
30. $\forall x\, \exists y\, xy \equiv e$	$\exists z\, xz \equiv e$	5.7(b3) applied to 29
31. φ_2	$\forall y\, \exists z\, yz \equiv e$	5.7(b2) applied to 30
32. $\varphi_0\ \varphi_1\ \varphi_2\ xy \equiv e$	$\exists y\, yx \equiv e$	(Ant), (Ch) applied to 31 and 26
33. $\varphi_0\ \varphi_1\ \varphi_2\ \forall x\, \exists y\, xy \equiv e$	$\exists y\, yx \equiv e$	($\exists$A) and 5.7(b3) applied to 32
34. $\varphi_0\ \varphi_1\ \varphi_2$	$\forall x\, \exists y\, yx \equiv e$	(Ant) and 5.7(b4) applied to 33

6.3 Exercise. Following the proof of I.2.1, give a derivation for the sequent

$$\psi_0\ \psi_1\ \psi_2\, \forall x\, \forall y\, (\exists u (Rxu \wedge Ryu) \rightarrow \forall z\, (Rxz \leftrightarrow Ryz)),$$

where ψ_0, ψ_1, and ψ_2 are the axioms for equivalence relations (cf. III.6.1).

§7. Consistency

The syntactic concept $\vdash$ of derivability corresponds to the semantic concept $\models$ of consequence. As a syntactic counterpart to satisfiability we define the concept of *consistency*.

7.1 Definition. (a) Φ is *consistent* (written: Con Φ) if and only if there is no formula φ such that $\Phi \vdash \varphi$ and $\Phi \vdash \neg\varphi$.
(b) Φ is *inconsistent* (written: Inc Φ) if and only if Φ is not consistent (that is, if there is a formula φ such that $\Phi \vdash \varphi$ and $\Phi \vdash \neg\varphi$).

First we show that from an inconsistent set one can derive any formula.

7.2 Lemma. *For a set of formulas Φ the following are equivalent:*

(a) Inc Φ,
(b) *For all φ, $\Phi \vdash \varphi$.*

PROOF. (a) follows immediately from (b). Suppose, on the other hand, that Inc Φ holds, i.e., $\Phi \vdash \psi$ and $\Phi \vdash \neg\psi$ for some formula ψ. Let φ be an arbitrary formula. We show $\Phi \vdash \varphi$. First of all there exist Γ_1 and Γ_2 consisting of formulas from Φ and derivations

$$\begin{array}{ll} \vdots & \\ \Gamma_1 & \psi \end{array} \quad \text{and} \quad \begin{array}{ll} \equiv & \\ \Gamma_2 & \neg\psi \end{array}.$$

By combining these two derivations and adding to the resulting derivation, we obtain

$$\begin{array}{llll} & \vdots & & \\ m. & \Gamma_1 & \psi & \\ & \equiv & & \\ n. & \Gamma_2 & \neg\psi & \\ n+1. & \Gamma_1\Gamma_2 & \psi & \text{(Ant) applied to } m \\ n+2. & \Gamma_1\Gamma_2 & \neg\psi & \text{(Ant) applied to } n \\ n+3. & \Gamma_1\Gamma_2 & \varphi & \text{(Ctr') applied to } n+1 \text{ and } n+2. \end{array}$$

Thus we see that $\Phi \vdash \varphi$. □

7.3 Corollary. *For a set of formulas* Φ *the following are equivalent:*

(a) Con Φ
(b) *There is a formula* φ *which is not derivable from* Φ. □

Since $\Phi \vdash \varphi$ if and only if $\Phi_0 \vdash \varphi$ for a suitable finite subset Φ_0 of Φ, we obtain:

7.4 Lemma. *For all* Φ, Con Φ *if and only if* Con Φ_0 *for all* finite *subsets* Φ_0 *of* Φ. □

7.5 Lemma. *Every satisfiable set of formulas is consistent.*

PROOF. Suppose Inc Φ. Then for a suitable φ both $\Phi \vdash \varphi$ and $\Phi \vdash \neg\varphi$; hence, by the theorem on the correctness of $\mathfrak{S}$, $\Phi \models \varphi$ and $\Phi \models \neg\varphi$. But then Φ cannot be satisfiable. □

Later we shall need:

7.6 Lemma. *For all* Φ *and* φ,

(a) *if not* $\Phi \vdash \varphi$, *then* Con $\Phi \cup \{\neg\varphi\}$;
(b) *if* Con Φ *and* $\Phi \vdash \varphi$, *then* Con $\Phi \cup \{\varphi\}$;
(c) *if* Con Φ, *then* Con $\Phi \cup \{\varphi\}$ *or* Con $\Phi \cup \{\neg\varphi\}$.

Proof. (a) Suppose not $\Phi \vdash \varphi$, but $\Phi \cup \{\neg \varphi\}$ is inconsistent. Then for a suitable Γ consisting of formulas from Φ, there is a derivation of the sequent $\Gamma \neg \varphi \varphi$. From this we obtain the following derivation:

$$\begin{array}{llll} & \vdots & & \\ \Gamma & \neg\varphi & \varphi & \\ \Gamma & \varphi & \varphi & \text{(Ass)} \\ \Gamma & & \varphi & \text{(PC)} \end{array}$$

Hence $\Phi \vdash \varphi$, a contradiction.

(b) Interchange the rôles of φ and $\neg \varphi$ in (a) and note that $\Phi \vdash \neg \varphi$ does not hold.

(c) This follows directly from (a) and (b). □

In this chapter we have referred to a fixed symbol set S. Thus when we spoke of formulas we understood them to be S-formulas, and when discussing the sequent calculus $\mathfrak{S}$ we actually referred to the particular calculus $\mathfrak{S}_S$ corresponding to the symbol set S. In some cases it is necessary to treat several symbol sets simultaneously. Then we insert indices for the sake of clarity. To be specific, we use the more precise notation $\Phi \vdash_S \varphi$ to indicate that there is a derivation in $\mathfrak{S}_S$ (consisting of S-formulas) whose last sequent is of the form $\Gamma \varphi$, where Γ consists of formulas from Φ. Similarly, we write $\mathrm{Con}_S\, \Phi$ if there is no S-formula φ such that $\Phi \vdash_S \varphi$ and $\Phi \vdash_S \neg \varphi$.[2]

In the next chapter we shall need:

7.7 Lemma. *For $n \in \mathbb{N}$, let S_n be symbol sets such that $S_0 \subset S_1 \subset S_2 \subset \ldots$, and let Φ_n be sets of S_n-formulas such that* $\mathrm{Con}_{S_n}\, \Phi_n$ *and $\Phi_0 \subset \Phi_1 \subset \Phi_2 \subset \ldots$. Let $S = \bigcup_{n \in \mathbb{N}} S_n$ and $\Phi = \bigcup_{n \in \mathbb{N}} \Phi_n$. Then* $\mathrm{Con}_S\, \Phi$.

Proof. Assume the hypotheses of the theorem, and suppose $\mathrm{Inc}_S\, \Phi$. Then, by 7.4, $\mathrm{Inc}_S\, \Psi$ must hold for a suitable *finite* subset Ψ of Φ. There is a k such that $\Psi \subset \Phi_k$ and hence $\mathrm{Inc}_S\, \Phi_k$; in particular, $\Phi_k \vdash_S v_0 \equiv v_0$ and $\Phi_k \vdash_S \neg v_0 \equiv v_0$. Suppose we are given S-derivations for these two formulas. Since they contain only a finite number of symbols, all the formulas occurring there are actually contained in some L^{S_m}. We may assume that $m \geq k$. Then both derivations are derivations in the S_m-sequent calculus and therefore $\mathrm{Inc}_{S_m}\, \Phi_k$. Since $\Phi_k \subset \Phi_m$ we then obtain $\mathrm{Inc}_{S_m}\, \Phi_m$, which contradicts the hypotheses of the theorem. □

[2] The reader should note that for two symbol sets $S' \supset S$, and for $\Phi \subset L^S$ and $\varphi \in L^S$, it is conceivable that $\Phi \vdash_{S'} \varphi$ but not $\Phi \vdash_S \varphi$, for it could be that formulas from $L^{S'} - L^S$ are used in *every* derivation of φ from Φ in $\mathfrak{S}_{S'}$, and that (later on in the proof) these formulas are then eliminated from the sequents, say by application of the rules (Ctr), (PC), or ($\exists$S). In V.4. we shall show that this cannot, in fact, happen.

7.8 Exercise. Define $(\exists\forall)$ to be the rule

$$\overline{\Gamma \; \exists x\varphi \; \forall x\varphi}.$$

(a) Determine whether $(\exists\forall)$ is a derivable rule.
(b) Let $\mathfrak{S}'$ be obtained from the calculus of sequents $\mathfrak{S}$ by adding the rule $(\exists\forall)$. Is every sequent derivable in $\mathfrak{S}'$?

CHAPTER V
The Completeness Theorem

The subject of this chapter is a proof of the completeness of the sequent calculus, i.e., the statement:

(*) For all Φ and φ, if $\Phi \models \varphi$ then $\Phi \vdash \varphi$.

In order to verify (*) we show

(**) Every consistent set of formulas is satisfiable.

From this, (*) can be proved as follows: We assume for Φ and φ that $\Phi \models \varphi$, but not $\Phi \vdash \varphi$. Then $\Phi \cup \{\neg \varphi\}$ is consistent (cf. Chapter IV, 7.6(a)) but not satisfiable, a contradiction to (**).

To establish (**) we have to find a model for any consistent set Φ of formulas. In §1 we shall see that there is a natural way to do this if Φ is *maximally consistent* and if it *contains witnesses*. Then we reduce the general case to this one: In §2 for at most countable symbol sets, and in §3 for arbitrary symbol sets.

Unless stated otherwise, we refer to a fixed symbol set S.

§1. Henkin's Theorem

Let Φ be a consistent set of formulas. In order to find a model $\mathfrak{I} = (\mathfrak{A}, \beta)$ of Φ, one can only use the "syntactical" information given by the consistency of Φ. Hence we shall try to obtain a model using syntactical objects as far as possible. A first idea is to take as domain A the set of all S-terms, to define β by $\beta(v_i) = v_i \, (i \in \mathbb{N})$ and $R^{\mathfrak{A}}$, say for unary R, by $R^{\mathfrak{A}} = \{t \in A \mid Rt \in \Phi\}$.

Then, if for instance $Rx \in \Phi$ and $Rx \to Ry \in \Phi$, we should have $Ry \in \Phi$, and if $\exists x\, Rx \in \Phi$ there should be a "witness" t, i.e. a term t such that $Rt \in \Phi$. We see that in order to get a model of Φ in this way, Φ has to satisfy certain closure conditions. These are made precise in the following definition. It will turn out that they are sufficient to carry out the above idea.

1.1 Definition. Let Φ be a set of formulas.

(a) Φ is said to be *maximally consistent* if and only if Con Φ, and if every formula φ with Con $\Phi \cup \{\varphi\}$ already belongs to Φ.

(b) Φ *contains witnesses* if and only if for every formula of the form $\exists x\varphi$ there exists a term t such that $\left(\exists x\varphi \to \varphi\frac{t}{x}\right) \in \Phi$.

If $\mathfrak{I}$ is an interpretation, then the set $\Phi = \{\varphi \in L^S \mid \mathfrak{I} \models \varphi\}$ is maximally consistent: Since $\mathfrak{I} \models \Phi$, Φ is satisfiable and hence by IV.7.5 it is consistent. Further, if Con $\Phi \cup \{\varphi\}$, then $\neg\varphi \notin \Phi$; hence $\mathfrak{I} \models \varphi$ and so $\varphi \in \Phi$.

Conversely, with every maximally consistent set Φ which contains witnesses we shall associate an interpretation $\mathfrak{I}_\Phi$ as outlined above such that $\mathfrak{I}_\Phi \models \Phi$. (Thus it turns out that every such Φ is satisfiable.)

1.2 Lemma. *Let Φ be maximally consistent and contain witnesses. Then, for all φ and ψ:*

(a) *If $\Phi \vdash \varphi$, then $\varphi \in \Phi$.*

(b) *Either $\varphi \in \Phi$ or $\neg\varphi \in \Phi$.*

(c) *$(\varphi \vee \psi) \in \Phi$ if and only if $\varphi \in \Phi$ or $\psi \in \Phi$.*

(d) *If $(\varphi \to \psi) \in \Phi$ and $\varphi \in \Phi$, then $\psi \in \Phi$.*

(e) *$\exists x\varphi \in \Phi$ if and only if there is a term t such that $\varphi\frac{t}{x} \in \Phi$.*

PROOF. (a) If $\Phi \vdash \varphi$, then by IV.7.6(b), Con $\Phi \cup \{\varphi\}$, and hence $\varphi \in \Phi$ since Φ is maximally consistent.

(b) By IV.7.6(c) we have Con $\Phi \cup \{\varphi\}$ or Con $\Phi \cup \{\neg\varphi\}$, and therefore $\varphi \in \Phi$ or $\neg\varphi \in \Phi$. Since $\Phi \cup \{\varphi, \neg\varphi\}$ is inconsistent, φ and $\neg\varphi$ cannot both belong to Φ.

(c) Suppose first that $(\varphi \vee \psi) \in \Phi$. If $\varphi \notin \Phi$ then $\neg\varphi \in \Phi$. Since $\vdash (\varphi \vee \psi)\, \neg\varphi\, \psi$ (cf. IV.3.6(b)), we have $\Phi \vdash \psi$ and from (a), $\psi \in \Phi$. On the other hand, if, for example, $\varphi \in \Phi$, then by IV.3.6(a1), $\Phi \vdash (\varphi \vee \psi)$, and so by (a), $(\varphi \vee \psi) \in \Phi$.

(d) Assume that $(\varphi \to \psi)$ (i.e. $(\neg\varphi \vee \psi)$) and φ belong to Φ. Since $\vdash (\neg\varphi \vee \psi)\varphi\psi$ (cf. IV.3.6(c)), we obtain by (a) that ψ belongs to Φ.

(e) First suppose that $\exists x\varphi \in \Phi$. Since Φ contains witnesses, there is a term t such that $\left(\exists x\varphi \to \varphi\frac{t}{x}\right) \in \Phi$, and therefore $\varphi\frac{t}{x} \in \Phi$ by (d). On the other hand, if $\varphi\frac{t}{x} \in \Phi$, then $\Phi \vdash \exists x\varphi$ (use ($\exists$S)) and by (a), $\exists x\varphi \in \Phi$. □

From 1.2 we shall obtain the result that for an interpretation $\mathfrak{I}$, the statement

$$(*) \qquad \mathfrak{I} \models \varphi \quad \text{iff } \varphi \in \Phi$$

holds for all φ (and hence $\mathfrak{I} \models \Phi$) provided we can establish (1) and (2):

(1) (*) holds for atomic φ.
(2) For every element of the domain of $\mathfrak{I}$ there is a term t such that $\mathfrak{I}(t) = a$.

Taking up our original idea we construct an interpretation $\mathfrak{I}$ satisfying (1) and (2). For the domain of $\mathfrak{I}$ we intended to take the set of terms, and to arrange the interpretation so that $\mathfrak{I}(t) = t$ (cf. (2)) and ($\mathfrak{I} \models Rt_0 \ldots t_{n-1}$ iff $Rt_0 \ldots t_{n-1} \in \Phi$) (cf. (1)). A slight difficulty arises concerning equations: If $t_0 \equiv t_1 \in \Phi$, then on account of (*), $\mathfrak{I}(t_0) = \mathfrak{I}(t_1)$ must hold even if t_0 and t_1 are distinct terms. We overcome this difficulty by defining an equivalence relation on terms and then using the equivalence classes rather than the individual terms as elements of the domain of $\mathfrak{I}$.

In the remainder of this section let Φ be a maximally consistent set containing witnesses. We proceed to define the interpretation $\mathfrak{I}_\Phi = (\mathfrak{T}_\Phi, \beta_\Phi)$. First of all, we introduce a binary relation $\sim$ on the set T^S of S-terms:

1.3. $t_0 \sim t_1$ *iff* $t_0 \equiv t_1 \in \Phi$.

1.4 Lemma. (a) $\sim$ *is an equivalence relation.*

(b) $\sim$ *is compatible with the symbols in S in the following sense: If $t_0 \sim t'_0, \ldots, t_{n-1} \sim t'_{n-1}$, then for n-ary $f \in S$,*

$$ft_0 \ldots t_{n-1} \sim ft'_0 \ldots t'_{n-1}$$

and for n-ary $R \in S$,

$$Rt_0 \ldots t_{n-1} \in \Phi \quad \textit{iff} \quad Rt'_0 \ldots t'_{n-1} \in \Phi.$$

The proof uses the rule ($\equiv$) and IV.5.3–5.6. We give two cases as examples:

(1) $\sim$ is symmetric: Suppose $t_0 \sim t_1$, that is, $t_0 \equiv t_1 \in \Phi$. By IV.5.3 we obtain $\Phi \vdash t_1 \equiv t_0$; hence, by 1.2(a), $t_1 \equiv t_0 \in \Phi$, i.e., $t_1 \sim t_0$.
(2) Let f be an n-ary function symbol from S, and assume $t_0 \sim t'_0, \ldots, t_{n-1} \sim t'_{n-1}$, i.e., $t_0 \equiv t'_0 \in \Phi, \ldots, t_{n-1} \equiv t'_{n-1} \in \Phi$. Then by IV.5.6, $\Phi \vdash ft_0 \ldots t_{n-1} \equiv ft'_0 \ldots t'_{n-1}$, and by 1.2(a), $ft_0 \ldots t_{n-1} \sim ft'_0 \ldots t'_{n-1}$. □

Let $\bar{t}$ be the equivalence class of t,

$$\bar{t} := \{t' \in T^S \mid t \sim t'\},$$

and let T_Φ be the set of equivalence classes,

$$T_\Phi := \{\bar{t} \mid t \in T^S\}.$$

T_Φ is not empty. Define the S-structure $\mathfrak{T}_\Phi$ over T_Φ, the so-called *term structure* corresponding to Φ, by the following clauses:

1.5. For n-ary $R \in S$,

$$R^{\mathfrak{T}_\Phi}\bar{t}_0 \ldots \bar{t}_{n-1} \quad \text{iff} \quad Rt_0 \ldots t_{n-1} \in \Phi.$$

1.6. For n-ary $f \in S$,

$$f^{\mathfrak{T}_\Phi}(\bar{t}_0, \ldots, \bar{t}_{n-1}) := \overline{ft_0 \ldots t_{n-1}}.$$

1.7. For $c \in S$,

$$c^{\mathfrak{T}_\Phi} := \bar{c}.$$

By 1.4 the conditions in 1.5 and 1.6 are independent of the choice of the representatives $t_0, \ldots, t_{n-1}$, hence $R^{\mathfrak{T}_\Phi}$ and $f^{\mathfrak{T}_\Phi}$ are well defined.

Finally, we fix an assignment β_Φ by

1.8. $\beta_\Phi(x) := \bar{x}$.

We call $\mathfrak{I}_\Phi = (\mathfrak{T}_\Phi, \beta_\Phi)$ the *term interpretation* associated with Φ.

1.9 Lemma. (a) *For all t, $\mathfrak{I}_\Phi(t) = \bar{t}$.*
(b) *For every atomic formula φ,*

$$\mathfrak{I}_\Phi \models \varphi \quad \textit{iff} \quad \varphi \in \Phi.$$

PROOF. (a) By induction on terms. The lemma holds for $t = x$ by 1.8 and for $t = c$ by 1.7. If $t = ft_0 \ldots t_{n-1}$, then

$$\begin{aligned} \mathfrak{I}_\Phi(ft_0 \ldots t_{n-1}) &= f^{\mathfrak{T}_\Phi}(\mathfrak{I}_\Phi(t_0), \ldots, \mathfrak{I}_\Phi(t_{n-1})) \\ &= f^{\mathfrak{T}_\Phi}(\bar{t}_0, \ldots, \bar{t}_{n-1}) \quad \text{(by induction hypothesis)} \\ &= \overline{ft_0 \ldots t_{n-1}} \quad \text{(by 1.6).} \end{aligned}$$

(b)

$$\begin{aligned} \mathfrak{I}_\Phi \models t_0 \equiv t_1 \quad &\text{iff} \quad \mathfrak{I}_\Phi(t_0) = \mathfrak{I}_\Phi(t_1) \\ &\text{iff} \quad \bar{t}_0 = \bar{t}_1 \quad \text{(by (a))} \\ &\text{iff} \quad t_0 \sim t_1 \\ &\text{iff} \quad t_0 \equiv t_1 \in \Phi. \end{aligned}$$

$$\begin{aligned} \mathfrak{I}_\Phi \models Rt_0 \ldots t_{n-1} \quad &\text{iff} \quad R^{\mathfrak{T}_\Phi}\bar{t}_0 \ldots \bar{t}_{n-1} \\ &\text{iff} \quad Rt_0 \ldots t_{n-1} \in \Phi \quad \text{(by 1.5).} \end{aligned}$$

□

1.10 Henkin's Theorem. *Let Φ be a maximally consistent set containing witnesses. Then for all φ,*

$$(*) \qquad \mathfrak{I}_\Phi \models \varphi \quad \textit{iff} \quad \varphi \in \Phi.$$

PROOF. We show (∗) by induction on the number of connectives and quantifiers in φ, in other words, by induction on $\text{rk}(\varphi)$ (cf. III.8.6). If $\text{rk}(\varphi) = 0$, then φ is atomic, and (∗) holds by 1.9(b). The induction step splits into three separate cases.

(1) $\varphi = \neg\psi$:

$$\begin{aligned} \mathfrak{I}_\Phi \models \neg\psi \quad & \text{iff not } \mathfrak{I}_\Phi \models \psi \\ & \text{iff } \psi \notin \Phi && \text{(by induction hypothesis)} \\ & \text{iff } \neg\psi \in \Phi && \text{(by 1.2(b)).} \end{aligned}$$

(2) $\varphi = (\psi \vee \chi)$:

$$\begin{aligned} \mathfrak{I}_\Phi \models (\psi \vee \chi) \quad & \text{iff } \mathfrak{I}_\Phi \models \psi \text{ or } \mathfrak{I}_\Phi \models \chi \\ & \text{iff } \psi \in \Phi \text{ or } \chi \in \Phi && \text{(by induction hypothesis)} \\ & \text{iff } (\psi \vee \chi) \in \Phi && \text{(by 1.2(c)).} \end{aligned}$$

(3) $\varphi = \exists x\psi$:

$\mathfrak{I}_\Phi \models \exists x\psi$

iff there is a term t such that $\mathfrak{I}_\Phi \dfrac{\bar{t}}{x} \models \psi$

iff there is a term t such that $\mathfrak{I}_\Phi \dfrac{\mathfrak{I}_\Phi(t)}{x} \models \psi$ (by 1.9(a))

iff there is a term t such that $\mathfrak{I}_\Phi \models \psi \dfrac{t}{x}$ (by the substitution lemma)

iff there is a term t such that $\psi \dfrac{t}{x} \in \Phi$

(by induction hypothesis since $\text{rk}\left(\psi \dfrac{t}{x}\right) = \text{rk}(\psi) < \text{rk}(\varphi)$ (cf. III.8.7))

iff $\exists x\psi \in \Phi$ (by 1.2(e)). □

1.11 Corollary. *Let Φ be a maximally consistent set containing witnesses. Then $\mathfrak{I}_\Phi \models \Phi$ and therefore Φ is satisfiable.* □

§2. Satisfiability of Consistent Sets of Formulas (the Countable Case)

By 1.11, every maximally consistent set of formulas containing witnesses is satisfiable. We prove that any consistent set Φ of formulas is satisfiable by showing how to extend it to a maximally consistent set containing witnesses. In this section we settle the case of symbol sets which are at most countable.

In the following let S be at most countable. First we treat the case where only finitely many variables occur free in Φ, i.e., where $\text{free}(\Phi) := \bigcup_{\varphi \in \Phi} \text{free}(\varphi)$ is finite.

2.1 Lemma. *Let $\Phi \subset L^S$ be consistent and let* free(Φ) *be finite. Then there is a consistent set Ψ such that $\Phi \subset \Psi \subset L^S$ and Ψ contains witnesses.*

2.2 Lemma. *Let $\Psi \subset L^S$ be consistent. Then there is a maximally consistent set Θ with $\Psi \subset \Theta \subset L^S$.*

2.1 and 2.2 enable us to extend a consistent set Φ of formulas in two stages to a maximally consistent set of formulas containing witnesses. First of all, we extend Φ to Ψ according to 2.1, and then Ψ to Θ according to 2.2. Θ is maximally consistent, and it contains witnesses because Ψ does. Hence by 1.11, Θ is satisfiable, and, since $\Phi \subset \Theta$, Φ is also satisfiable. To summarize we obtain:

2.3 Corollary. *Let Φ be consistent, and let* free(Φ) *be finite. Then Φ is satisfiable.* □

It still remains to prove 2.1 and 2.2.

Proof of Lemma 2.1. By II.3.3, L^S is countable. Let $\exists x_0 \varphi_0, \exists x_1 \varphi_1, \ldots$ be a list of all formulas in L^S which begin with an existential quantifier. Inductively we define formulas $\psi_0, \psi_1, \ldots$, which we add to Φ. For each n, ψ_n is a "witness formula" for $\exists x_n \varphi_n$.

Suppose ψ_m is already defined for $m < n$. Since free(Φ) is finite, only finitely many variables occur free in $\Phi \cup \{\psi_m \mid m < n\} \cup \{\exists x_n \varphi_n\}$. Let y_n be a variable distinct from these. We set

$$\psi_n := \left(\exists x_n \varphi_n \rightarrow \varphi_n \frac{y_n}{x_n}\right).$$

Now let

$$\Psi := \Phi \cup \{\psi_0, \psi_1, \ldots\}.$$

Then $\Phi \subset \Psi$ and Ψ clearly contains witnesses. It remains to be shown that Ψ is consistent. For this purpose put

$$\Phi_n := \Phi \cup \{\psi_m \mid m < n\}.$$

Then $\Phi_0 \subset \Phi_1 \subset \Phi_2 \subset \cdots$ and $\Psi = \bigcup_{n \in \mathbb{N}} \Phi_n$. By IV.7.7 (for $S = S_0 = S_1 = \ldots$) the proof will be complete if we can show that each Φ_n is consistent. We proceed by induction on n.

Since $\Phi_0 = \Phi$, Con Φ_0 holds by hypothesis. For the induction step we assume that Φ_n is consistent. Suppose, for a contradiction, that $\Phi_{n+1} = \Phi_n \cup \{\psi_n\}$ is inconsistent. Then for every φ there exists Γ over Φ_n such that $\vdash \Gamma\ \psi_n \varphi$, i.e.,

$$\vdash \Gamma\left(\neg \exists x_n \varphi_n \vee \varphi_n \frac{y_n}{x_n}\right)\varphi.$$

Thus there is a derivation,

$$\vdots$$

$$m.\quad \Gamma\left(\neg \exists x_n \varphi_n \vee \varphi_n \frac{y_n}{x_n}\right)\varphi,$$

which, if φ is a sentence, we can extend as follows:

$m+1.$	$\Gamma\ \neg\exists x_n\varphi_n\ \neg\exists x_n\varphi_n$	(Ass)
$m+2.$	$\Gamma\ \neg\exists x_n\varphi_n\ \left(\neg\exists x_n\varphi_n \vee \varphi_n \frac{y_n}{x_n}\right)$	$(\vee S)$ applied to $m+1$
$m+3.$	$\Gamma\ \neg\exists x_n\varphi_n\ \varphi$	(Ch) (with (Ant)) applied to $m+2$ and m
$m+4.$	$\Gamma\ \varphi_n \frac{y_n}{x_n}\ \varphi$	(analogously)
$m+5.$	$\Gamma\ \exists x_n\varphi_n\ \varphi$	$(\exists A)$ applied to $m+4$ (y_n does not occur free in $\Gamma\ \exists x_n\varphi_n\ \varphi$)
$m+6.$	$\Gamma\ \varphi$	(PC) applied to $m+5$ and $m+3$

Hence we have $\Phi_n \vdash \varphi$. But then Φ_n is inconsistent, as can be seen by taking $\varphi = \exists v_0\, v_0 \equiv v_0$ and $\varphi = \neg\exists v_0\, v_0 \equiv v_0$. This contradicts the induction hypothesis. □

Proof of Lemma 2.2. Suppose Ψ is consistent and let $\varphi_0, \varphi_1, \varphi_2, \ldots$ be an enumeration of L^S. We define sets of formulas Θ_n inductively as follows:

$$\Theta_0 := \Psi,$$

and

$$\Theta_{n+1} := \begin{cases} \Theta_n \cup \{\varphi_n\} & \text{if Con } \Theta_n \cup \{\varphi_n\} \\ \Theta_n & \text{otherwise,} \end{cases}$$

and we set

$$\Theta := \bigcup_{n\in\mathbb{N}} \Theta_n.$$

First of all, $\Psi \subset \Theta$. Clearly all Θ_n are consistent, and hence by IV.7.7, Θ is consistent as well. Finally, Θ is maximally consistent. For if $\varphi \in L^S$, say $\varphi = \varphi_n$, and if Con $\Theta \cup \{\varphi_n\}$, then, since $\Theta_n \subset \Theta$, we obtain Con $\Theta_n \cup \{\varphi_n\}$ and hence $\varphi_n \in \Theta_{n+1}$, i.e., $\varphi_n \in \Theta$. □

Now we drop the assumption that free(Φ) is finite.

2.4 Theorem. *If S is at most countable and $\Phi \subset L^S$ is consistent, then Φ is satisfiable.*

Proof. We reduce 2.4 to 2.3. Let $c_0, c_1, \ldots$ be distinct constants which do not belong to S, and set

$$S' = S \cup \{c_0, c_1, \ldots\}.$$

For $\varphi \in L^S$ denote by $n(\varphi)$ the smallest n such that free(φ) $\subset \{v_0, \ldots, v_{n-1}\}$. Let

$$(1) \qquad \varphi' := \varphi \frac{c_0 \ldots c_{n(\varphi)-1}}{v_0 \ldots v_{n(\varphi)-1}} \quad \text{and} \quad \Phi' := \{\varphi' \mid \varphi \in \Phi\}.$$

First, by III.8.5,

$$(2) \qquad \text{free}(\Phi') = \varnothing.$$

Now it will suffice to show that

$$(3) \qquad \text{Con}_{S'} \Phi',$$

for then we know from the special case proved in 2.3 that Φ' is satisfiable, say by some interpretation $\mathfrak{I}' = (\mathfrak{A}', \beta')$. Using the coincidence lemma we can, by (2), assume that $\beta'(v_n) = c_n^{A'}$, i.e., $\mathfrak{I}'(v_n) = \mathfrak{I}'(c_n)$ for all $n \in \mathbb{N}$. Then from (1) and the substitution lemma it follows that $\mathfrak{I}'$ is a model of Φ; hence Φ is satisfiable.

We prove (3) by showing that every finite subset Φ'_0 of Φ' is satisfiable, and thus, by IV.7.5, consistent (with respect to S'). Let $\Phi'_0 = \{\varphi'_0, \ldots, \varphi'_{n-1}\}$, where $\varphi_0, \ldots, \varphi_{n-1} \in \Phi$. Since $\{\varphi_0, \ldots, \varphi_{n-1}\}$ is a subset of Φ it is consistent (with respect to S), and since only finitely many variables occur free therein, it is satisfiable (cf. 2.3).

Choose an S-interpretation $\mathfrak{I} = (\mathfrak{A}, \beta)$ such that

$$(*) \qquad \mathfrak{I} \models \{\varphi_0, \ldots, \varphi_{n-1}\}$$

and expand $\mathfrak{A}$ to an S'-structure $\mathfrak{A}'$ with $c_n^{\mathfrak{A}'} = \mathfrak{I}(v_n)$ for $n \in \mathbb{N}$. From (1), (*), and the substitution lemma, it then follows that the S'-interpretation $(\mathfrak{A}', \beta')$ which results is a model of Φ'_0. □

The following exercise shows that the assumption "free(Φ) is finite" in 2.1 is necessary.

2.5 Exercise. Let S be arbitrary and let

$$\Phi = \{v_0 \equiv t \mid t \in T^S\} \cup \{\exists v_0 \exists v_1 \neg v_0 \equiv v_1\}.$$

Show that Con Φ holds and that there is no consistent set in L^S which includes Φ and contains witnesses.

2.6 Exercise (A Special Case of the So-called *Herbrand Theorem*). Let S be a symbol set, and let φ and ψ be quantifier-free formulas in which there is at most one free variable, namely x. Show that if

$$\forall x\varphi \models \exists x\psi,$$

then there is an $n \geq 1$ and there are S-terms $t_0, \ldots, t_{n-1}, s_0, \ldots, s_{n-1}$ such that

$$\varphi\frac{t_0}{x} \wedge \cdots \wedge \varphi\frac{t_{n-1}}{x} \models \psi\frac{s_0}{x} \vee \cdots \vee \psi\frac{s_{n-1}}{x}.$$

(*Hint*: Give an indirect proof making use of the set

$$\left\{\varphi\frac{t}{x} \,\middle|\, t \text{ is an } S\text{-term}\right\} \cup \left\{\neg\psi\frac{t}{x} \,\middle|\, t \text{ is an } S\text{-term}\right\}.)$$

§3. Satisfiability of Consistent Sets of Formulas (the General Case)

In this section we no longer assume that S is countable. The rôle of 2.1 and 2.2 will be taken over by 3.1 and 3.2.

3.1 Lemma. *Assume* $\Phi \subset L^S$ *and* $\mathrm{Con}_S\,\Phi$. *Then there is an* $S' \supset S$ *and a* Ψ *such that* $\Phi \subset \Psi \subset L^{S'}$ *and* $\mathrm{Con}_{S'}\,\Psi$, *and* Ψ *contains witnesses with respect to* S' *(i.e., for every formula of the form* $\exists x\varphi \in L^{S'}$ *there is a term* $t \in T^{S'}$ *such that* $\left(\exists x\varphi \to \varphi\frac{t}{x}\right) \in \Psi$).

3.2 Lemma. *Assume* $\Psi \subset L^S$ *and* $\mathrm{Con}_S\,\Psi$. *Then there is a set* Θ *such that* $\Psi \subset \Theta \subset L^S$ *and* Θ *is maximally consistent with respect to* S.

We obtained 2.3 from 2.1 and 2.2; likewise we have from 3.1 and 3.2 the following:

3.3 Corollary. *If* $\Phi \subset L^S$ *and* $\mathrm{Con}_S\,\Phi$, *then* Φ *is satisfiable.* □

The following consideration will lead to a proof of 3.1.

Let S be an arbitrary symbol set. Associate with every $\varphi \in L^S$ a constant c_φ such that $c_\varphi \notin S$ and $c_\varphi \neq c_\psi$ for $\varphi \neq \psi$. Defining

$$S^* := S \cup \{c_{\exists x\varphi} \mid \exists x\varphi \in L^S\}$$

and

$$W(S) := \left\{\exists x\varphi \to \varphi\frac{c_{\exists x\varphi}}{x} \,\middle|\, \exists x\varphi \in L^S\right\},$$

one obtains for $\Phi \subset L^S$:

3.4. *If* $\mathrm{Con}_S\,\Phi$, *then* $\mathrm{Con}_{S^*}\,\Phi \cup W(S)$.

Proof. Suppose $\mathrm{Con}_S\,\Phi$ holds. We show that every finite subset Φ_0^* of $\Phi \cup W(S)$ is consistent with respect to S^* by proving that it is satisfiable.

Let

$$\Phi_0^* = \Phi_0 \cup \left\{\exists x_0\varphi_0 \to \varphi_0\frac{c_0}{x_0}, \ldots, \exists x_{n-1}\varphi_{n-1} \to \varphi_{n-1}\frac{c_{n-1}}{x_{n-1}}\right\},$$

where $\Phi_0 \subset \Phi$, $\exists x_0\varphi_0, \ldots, \exists x_{n-1}\varphi_{n-1} \in L^S$, and where c_i stands for $c_{\exists x_i\varphi_i}$.

For a suitable finite subset $S_0 \subset S$, we have

$$\Phi_0 \cup \{\exists x\varphi_0, \ldots, \exists x_{n-1}\varphi_{n-1}\} \subset L^{S_0}.$$

Further, since $\mathrm{Con}_S\,\Phi$ holds, so does $\mathrm{Con}_S\,\Phi_0$, and hence, of course, $\mathrm{Con}_{S_0}\,\Phi_0$. Because $\mathrm{free}(\Phi_0)$ is finite, it follows from 2.3 that Φ_0 is satisfiable.

Let $\mathfrak{I} = (\mathfrak{A}, \beta)$ be an S-interpretation which satisfies Φ_0 and fix an element a in A. In order to satisfy Φ_0^* we extend $\mathfrak{I}$ to an S^*-interpretation $\mathfrak{I}^*$ as follows: For $i < n$ we choose $a_i \in A$ such that

$$(*) \qquad \mathfrak{I}\frac{a_i}{x_i} \models \varphi_i, \text{ if } \mathfrak{I} \models \exists x_i\varphi_i,$$

and $a_i = a$ otherwise. We extend $\mathfrak{A}$ to an S^*-structure $\mathfrak{A}^*$ by setting

$$c_i^{\mathfrak{A}^*} = a_i$$

for $i < n$ and interpreting the remaining constants of the form $c_{\exists x\varphi}$ by a. Let $\mathfrak{I}^* = (\mathfrak{A}^*, \beta)$. Since no constant $c_{\exists x\varphi}$ occurs in Φ_0, it follows from $\mathfrak{I} \models \Phi_0$ that $\mathfrak{I}^* \models \Phi_0$. Furthermore

$$\mathfrak{I}^* \models \exists x_i\varphi_i \to \varphi_i\frac{c_i}{x_i}$$

(and this shows that Φ_0^* is satisfiable). In fact, if $\mathfrak{I}^* \models \exists x_i\varphi_i$ then $\mathfrak{I}^*\frac{a_i}{x_i} \models \varphi_i$ by $(*)$. Since $a_i = \mathfrak{I}^*(c_i)$ it follows by the substitution lemma that $\mathfrak{I}^* \models \varphi_i\frac{c_i}{x_i}$.

Proof of Lemma 3.1. Let $\Phi \subset L^S$ and suppose $\mathrm{Con}_S\,\Phi$. We define a symbol set S' and $\Psi \subset L^{S'}$ with the following properties:

(a) $S \subset S'$ and $\Phi \subset \Psi$.
(b) $\mathrm{Con}_{S'}\,\Psi$.
(c) Ψ contains witnesses.

For this purpose we define symbol sets S_n and sets Φ_n of formulas by induction on n:

$$S_0 := S \quad\text{and}\quad S_{n+1} := (S_n)^*.$$

$$\Phi_0 := \Phi \quad\text{and}\quad \Phi_{n+1} := \Phi_n \cup W(S_n).$$

(Concerning the definition of $(S_n)^*$ and $W(S_n)$, cf. the definitions of S^* and $W(S)$ preceding 3.4.)

From the construction it follows that

$$S = S_0 \subset S_1 \subset S_2 \subset \ldots,$$

$$\Phi_n \subset L^{S_n} \quad \text{for } n \in \mathbb{N},$$

$$\Phi = \Phi_0 \subset \Phi_1 \subset \Phi_2 \subset \ldots.$$

We set $S' := \bigcup_{n \in \mathbb{N}} S_n$ and $\Psi := \bigcup_{n \in \mathbb{N}} \Phi_n$. Then (a) holds. Using 3.4 one can easily show $\mathrm{Con}_{S_n} \Phi_n$ by induction on n, and hence by IV.7.7, that $\mathrm{Con}_{S'} \Psi$. Therefore (b) also holds. Finally, Ψ contains witnesses: Suppose, for example, $\exists x \varphi \in L^{S'}$. Then, for a suitable n, $\exists x \varphi \in L^{S_n}$. Thus for some constant $c \in S_{n+1}$, the formula $\left(\exists x \varphi \rightarrow \varphi \frac{c}{x}\right)$ is an element of $W(S_n)$ and hence an element of Ψ. □

PROOF OF LEMMA 3.2. In the proof of 2.2 we made essential use of the countability of L^S. For arbitrary S we no longer have this property at our disposal. We resort to *Zorn's lemma*, which we now state in a form suited to our purposes. The reader can find a proof of this lemma in books on set theory.

Let M be a set and let $\mathfrak{U}$ be a nonempty set of subsets of M. $\mathfrak{B}$ is called a *chain* in $\mathfrak{U}$, if $\mathfrak{B} \subset \mathfrak{U}$, $\mathfrak{B} \neq \emptyset$, and if for $V_0, V_1 \in \mathfrak{B}$ we have $V_0 \subset V_1$ or $V_1 \subset V_0$. Then Zorn's lemma says

3.5. If for every chain $\mathfrak{B}$ in $\mathfrak{U}$ the union $\bigcup_{V \in \mathfrak{B}} V$ belongs to $\mathfrak{U}$, then there is at least one maximal element in $\mathfrak{U}$, i.e., an element U_0 for which there is no $U_1 \in \mathfrak{U}$ such that $U_0 \subsetneqq U_1$.

Now, for arbitrary S, let $\Psi \subset L^S$ and $\mathrm{Con}\, \Psi$. Set $M := L^S$ and

$$\mathfrak{U} := \{\Phi \mid \Psi \subset \Phi \subset L^S \text{ and } \mathrm{Con}_S \Phi\}.$$

Clearly $\Psi \in \mathfrak{U}$, so $\mathfrak{U}$ is not empty. Let $\mathfrak{B}$ be a chain in $\mathfrak{U}$. $\Theta_1 := \bigcup_{\Phi \in \mathfrak{B}} \Phi$ is an element of $\mathfrak{U}$ because $\Psi \subset \Theta_1 \subset L^S$ and $\mathrm{Con}_S \Theta_1$. (The consistency of Θ_1 can be proved as follows: If Θ_0 is a finite subset of Θ_1, say $\Theta_0 = \{\varphi_0, \ldots, \varphi_{n-1}\}$, then there are $\Phi_0, \ldots, \Phi_{n-1} \in \mathfrak{B}$ with $\varphi_i \in \Phi_i$ for $i < n$. Since $\mathfrak{B}$ is a chain, we can number the Φ_i such that $\Phi_0 \subset \Phi_1 \subset \cdots \subset \Phi_{n-1}$. Thus $\Theta_0 \subset \Phi_{n-1}$, and by $\mathrm{Con}_S \Phi_{n-1}$ we have $\mathrm{Con}_S \Theta_0$.)

Now we can apply Zorn's lemma (3.5) to $\mathfrak{U}$, thereby obtaining a maximal element Θ in $\mathfrak{U}$. From the definition of $\mathfrak{U}$ we know that $\Psi \subset \Theta \subset L^S$ and $\mathrm{Con}_S \Theta$. On the other hand Θ is also maximally consistent. For if $\varphi \in L^S$ and $\mathrm{Con}_S \Theta \cup \{\varphi\}$, then $\Theta \cup \{\varphi\} \in \mathfrak{U}$; but since Θ is maximal, $\Theta = \Theta \cup \{\varphi\}$, in other words, $\varphi \in \Theta$. □

§4. The Completeness Theorem

As already mentioned in the introduction to this chapter, we can obtain the completeness of the sequent calculus from 2.3 (for at most countable S) and from 3.3 (for arbitrary S):

4.1 Completeness Theorem. *For* $\Phi \subset L^S$ *and* $\varphi \in L^S$, *if* $\Phi \models \varphi$, *then* $\Phi \vdash_S \varphi$. □

From 4.1 together with the theorem on correctness (IV.6.2) we have:

$$\text{For } \Phi \in L^S \quad \text{and} \quad \varphi \in L^S, \qquad \Phi \models \varphi \quad \text{iff } \Phi \vdash_S \varphi,$$

and from 3.3 and IV.7.5 we obtain:

$$\text{For } \Phi \subset L^S, \qquad \text{Sat } \Phi \quad \text{iff } \text{Con}_S \Phi.$$

In III.5 we saw that the concepts of consequence and of satisfiability are actually independent of the particular choice of S. It follows from the results above that the concepts of derivability and consistency are also independent of S (cf. the footnote on page 74). Thus we can simply write "$\vdash$" and "Con", omitting the subscript.

4.2 Theorem on the Adequacy of the Sequent Calculus.

(a) $\Phi \models \varphi$ *iff* $\Phi \vdash \varphi$.
(b) Sat Φ *iff* Con Φ. □

4.3 Exercise (cf. exercise III.2.1). If one transfers the rules (Ass), (Ant), (PC), (Ctr), ($\vee$A), and ($\vee$S) from the sequent calculus to the language of propositional calculus (with propositional variables $p_0, p_1, p_2, \ldots$ and connectives $\neg$ and $\vee$), one obtains a sequent calculus for propositional logic. For a set $\Phi \cup \{\alpha\}$ of formulas in propositional logic let $\Phi \vdash \alpha$ have a definition similar to the definition for first-order logic. Further, write $\Phi \models \alpha$ if for every assignment s such that $\beta[s] = \text{T}$ for all $\beta \in \Phi$, also $\alpha[s] = \text{T}$.

Prove the completeness theorem (and the correctness theorem) for propositional logic:

$$\Phi \models \alpha \quad \text{iff } \Phi \vdash \alpha.$$

Historical Note. The completeness theorem is due to Gödel [11]. The program of setting up a calculus of reasoning was first formulated and pursued by Leibniz, although traces of it may be found in the works of earlier philosophers (e.g., Aristotle and Lull). At the beginning of this century, Russell and Whitehead developed a calculus, and within it, gave formal proofs for a large number of mathematical theorems. Then in 1928, Gödel proved the completeness theorem. The method of proof used in this section is due to Henkin [13].

CHAPTER VI

The Löwenheim–Skolem Theorem and the Compactness Theorem

The equivalences of $\vdash$ and $\models$ and of Con and Sat, respectively, form a bridge between syntax and semantics which allows us to transfer properties of $\vdash$ to $\models$ and of Con to Sat. In this way we shall prove several important results concerning $\models$ and Sat, and at the same time we shall acquire a deeper insight into the expressive power of first-order languages.

§1. The Löwenheim–Skolem Theorem

The domain of the model $\mathfrak{I}_\Phi$ defined in V.1 consists of equivalence classes of terms. We use this fact to obtain the following theorem:

1.1 Löwenheim–Skolem Theorem. *Every satisfiable and at most countable set of formulas is satisfiable over a domain which is at most countable.*

Proof. First let Φ be an at most countable set of S-*sentences* which is satisfiable and hence consistent. Since each S-formula contains only finitely many S-symbols, there are at most countably many S-symbols in Φ. Therefore we may assume, without loss of generality, that S itself is at most countable. Since Sat Φ holds, so does Con Φ, and the proofs in V.1 and V.2 show that there is an interpretation which satisfies Φ and whose domain A consists of classes $\bar{t}$ of terms, where t ranges over T^S. Because T^S is countable (cf. II.3.3), A is at most countable. This argument can easily be transferred from sets of sentences to sets of formulas; for, if Φ is a set of S-formulas and

$$\Phi' = \left\{\psi \frac{c_0 \ldots c_{n-1}}{v_0 \ldots v_{n-1}} \middle| n \in \mathbb{N}, \psi \in L_n^S \cap \Phi\right\},$$

where $c_0, c_1, \ldots,$ are new constants, then Φ and Φ' are satisfiable over the same domains (cf. the proof of V.2.4). □

$\forall x \forall y\, x \equiv y$ is a sentence which has only finite models. For a unary function symbol f, the sentence $\forall x \forall y\, (fx \equiv fy \rightarrow x \equiv y) \wedge \neg \forall x \exists y\, fy \equiv x$ has only infinite models (since there is no function on a finite set which is injective but not surjective).

If one re-examines the proof of the completeness theorem for the case of uncountable symbol sets, one obtains the following generalization of 1.1, which we formulate for readers who are familiar with the concept of cardinality:

1.2 Downward Löwenheim–Skolem Theorem. *Every satisfiable set of formulas $\Phi \subset L^S$ is satisfiable over a domain of cardinality not greater than the cardinality of L^S.* □

In 1.1 (and 1.2) a certain weakness of first-order languages is already apparent. In the case of the symbol set $S_{\mathrm{ar}}^{<}$, for example, there cannot exist a set Φ of sentences which characterizes the ordered field $\mathfrak{R}^{<} = (\mathbb{R}, +, \cdot, 0, 1, <)$ up to isomorphism (in the sense that exactly $\mathfrak{R}^{<}$ and the structures isomorphic to $\mathfrak{R}^{<}$ are the models of Φ). Any such set Φ of $S_{\mathrm{ar}}^{<}$-sentences would be at most countable and satisfiable (since $\mathfrak{R}^{<} \models \Phi$ must hold); then by 1.1 there would be an at most countable structure $\mathfrak{A}$ such that $\mathfrak{A} \models \Phi$. But this could not be isomorphic to $\mathfrak{R}^{<}$ since $\mathbb{R}$ is uncountable.

In analysis $\mathfrak{R}^{<}$ is characterized up to isomorphism, say, by the axioms for ordered fields and the axiom on Dedekind cuts. Since the former can be formulated in $L^{S_{\mathrm{ar}}^{<}}$, we see that the axiom on Dedekind cuts cannot be phrased in terms of $S_{\mathrm{ar}}^{<}$-formulas.

1.3 Exercise. If Φ is an at most countable, satisfiable set of formulas and if the equality symbol does not occur in any formula of Φ, then Φ is satisfiable over a countable domain. (*Hint*: In the proof of Henkin's theorem use the set of terms instead of the set of classes of terms as the domain of $\mathfrak{I}_{\Phi}$.)

1.4 Exercise. Show that every at most countable set of formulas which is satisfiable over an infinite domain is satisfiable over a countable domain.

§2. The Compactness Theorem

From the definition of $\vdash$ and Con we obtained directly (cf. IV.6.1 and IV.7.4):

(a) $\Phi \vdash \varphi$ iff there is a finite $\Phi_0 \subset \Phi$ such that $\Phi_0 \vdash \varphi$.
(b) Con Φ iff for all finite $\Phi_0 \subset \Phi$, Con Φ_0.

Using the adequacy theorem V.4.2 we can rephrase these results for the corresponding semantic concepts:

2.1 Compactness Theorem

(a) (*for the consequence relation*)

$$\Phi \models \varphi \quad \textit{iff there is a finite } \Phi_0 \subset \Phi \textit{ such that } \Phi_0 \models \varphi.$$

(b) (*for satisfiability*)

$$\text{Sat } \Phi \quad \textit{iff for all finite } \Phi_0 \subset \Phi, \text{Sat } \Phi_0.$$

The compactness theorem is so called because, in terms of a suitable topological reformulation, it says that a certain topology is compact (cf. Exercise 2.5).

We now use the compactness theorem to obtain variants of the Löwenheim–Skolem theorem.

2.2 Theorem. *Let Φ be a set of formulas which is satisfiable over arbitrarily large finite domains (i.e., for every $n \in \mathbb{N}$ there is an interpretation satisfying Φ whose domain is finite and has at least n elements). Then Φ is also satisfiable over an infinite domain.*

PROOF. Let

$$\Psi := \Phi \cup \{\varphi_{\geq n} | 2 \leq n\}$$

($\varphi_{\geq n}$ was introduced in III.6.3). Every interpretation which satisfies Ψ is a model of Φ and has an infinite domain. Therefore we need only prove that Ψ is satisfiable. By the compactness theorem it is sufficient to show that every finite subset Ψ_0 of Ψ is satisfiable. For each such Ψ_0 there is an $n_0 \in \mathbb{N}$ such that

$$(*) \qquad \Psi_0 \subset \Phi \cup \{\varphi_{\geq n} | 2 \leq n \leq n_0\}.$$

According to the hypothesis of the theorem there is an interpretation $\mathfrak{I}$ which satisfies Φ and whose domain contains at least n_0 elements. By (*), $\mathfrak{I}$ is also a model of Ψ_0. □

2.3 Upward Löwenheim–Skolem Theorem. *Let Φ be a set of formulas which is satisfiable over an infinite domain. Then for every set A there is a model of Φ which contains at least as many elements as A. (We say that M has at least as many elements as A if there exists an injective map of A into M.)*

PROOF. Let $\Phi \subset L^S$. For each $a \in A$ let c_a be a new constant (i.e., $c_a \notin S$) such that $c_a \neq c_b$ for $a \neq b$. First, we show that the set

$$\Psi := \Phi \cup \{\neg c_a \equiv c_b | a, b \in A, a \neq b\}$$

of $S \cup \{c_a | a \in A\}$-formulas is satisfiable. Because of the compactness theorem we can restrict ourselves to showing, for every finite n-tuple of distinct elements $a_0, \ldots, a_{n-1} \in A$, that

$$(+) \qquad \Phi \cup \{\neg c_{a_i} \equiv c_{a_j} | i, j < n, i \neq j\}$$

is satisfiable (cf. the argument in the previous proof). By hypothesis, there is an S-interpretation $\mathfrak{I} = (\mathfrak{B}, \beta)$ which satisfies Φ and whose domain is infinite. Therefore there are n distinct elements $b_0, \ldots, b_{n-1}$ in B. We let $c_{a_i}^{\mathfrak{B}} := b_i$ for $i < n$. Then the interpretation $((\mathfrak{B}, c_{a_0}^{\mathfrak{B}}, \ldots, c_{a_{n-1}}^{\mathfrak{B}}), \beta)$ satisfies the set $(+)$. Since every finite subset of Ψ is satisfiable, we can find an interpretation $\mathfrak{I}'$ which satisfies Ψ and hence also Φ. Let D be the domain of $\mathfrak{I}'$. For $a, b \in A$ such that $a \neq b$ we have $\mathfrak{I}' \models \neg c_a \equiv c_b$. Hence $\mathfrak{I}'(c_a)$ and $\mathfrak{I}'(c_b)$ are distinct elements of D. Therefore the map $\pi: A \to D$, where $\pi(a) = \mathfrak{I}'(c_a)$, is injective. Thus D has at least as many elements as A. □

The same idea is used in the proof of the following theorem, which we state here for readers familiar with the concept of cardinality.

2.4 Theorem of Löwenheim, Skolem, and Tarski. *Let Φ be a set of formulas which is satisfiable over an infinite domain and let κ be an infinite cardinal greater than or equal to the cardinality of Φ. Then Φ has a model of cardinality κ.*

Proof. Let Φ and κ be given as in the statement of the theorem. Let A be a set of cardinality κ. We may assume that $\Phi \subset L^S$ for a symbol set S of cardinality $\leq \kappa$. Then the symbol set $S \cup \{c_a | a \in A\}$ given in the proof of 2.3 has cardinality κ as does the set of $S \cup \{c_a | a \in A\}$-formulas. Again, let $\Psi = \Phi \cup \{\neg c_a \equiv c_b | a, b \in A, a \neq b\}$. By 1.2 there is a model $\mathfrak{I}'$ of Ψ (and hence also of Φ) whose domain D has cardinality $\leq \kappa$. On the other hand, since $\neg c_a \equiv c_b \in \Psi$ for distinct $a, b \in A$, D has cardinality $\geq \kappa$; hence its cardinality is exactly κ. □

2.5 Exercise. Let S be a symbol set. For every satisfiable set Φ of S-sentences let $\mathfrak{A}_\Phi$ be an S-structure such that $\mathfrak{A}_\Phi \models \Phi$. Further, write

$$\Sigma := \{\mathfrak{A}_\Phi | \Phi \subset L_0^S, \text{Sat } \Phi\},$$

and for every S-sentence φ set $X_\varphi := \{\mathfrak{A} \in \Sigma | \mathfrak{A} \models \varphi\}$.

(a) Show that the system $\{X_\varphi | \varphi \in L_0^S\}$ is basis for a topology on Σ.
(b) Show that every set X_φ is closed.
(c) Use the compactness theorem to show that every open covering of Σ has a finite subcovering, so that Σ is (quasi-)compact.

§3. Elementary Classes

For a set Φ of S-sentences we call

$$\text{Mod}_S \Phi := \{\mathfrak{A} | \mathfrak{A} \text{ is an } S\text{-structure and } \mathfrak{A} \models \Phi\}$$

the *class of models* of Φ. Instead of "$\text{Mod}_S\{\varphi\}$" we sometimes write "$\text{Mod}_S \varphi$".

3.1 Definition. Let $\mathfrak{K}$ be a class of S-structures.

(a) $\mathfrak{K}$ is called *elementary* iff there is an S-sentence φ such that $\mathfrak{K} = \text{Mod}_S\, \varphi$.
(b) $\mathfrak{K}$ is called Δ-*elementary* iff there is a set Φ of S-sentences such that $\mathfrak{K} = \text{Mod}_S\, \Phi$.

Every elementary class is Δ-elementary. Conversely, because

$$\text{Mod}_S\, \Phi = \bigcap_{\varphi \in \Phi} \text{Mod}_S\, \varphi,$$

every Δ-elementary class is the intersection of elementary classes.

From an algebraic point of view we can formulate the question of the expressive power of first-order languages as follows: Which classes of structures are elementary or Δ-elementary, i.e., which classes can be axiomatized by a first-order sentence φ or by a set Φ of first-order sentences?

Let us give some examples.

3.2. *The class of fields* (considered as S_{ar}-structures) *and the class of ordered fields* (considered as $S_{\text{ar}}^{<}$-structures) *are elementary*. For example, the first class can be represented in the form $\text{Mod}_{S_{\text{ar}}}\, \varphi_F$, where φ_F is the conjunction of the field axioms in III.6.5. Similarly, the *class of groups*, the *class of equivalence structures* and the *class of partially defined orderings* (cf. III.6.4) are elementary.

Let p be a prime number. A field $\mathfrak{F}$ has characteristic p if

$$\underbrace{1^{\mathfrak{F}} + \cdots + 1^{\mathfrak{F}}}_{p\text{-times}} = 0^{\mathfrak{F}},$$

that is, if $\mathfrak{F}$ satisfies the sentence

$$\chi_p := \underbrace{1 + \cdots + 1}_{p\text{-times}} \equiv 0.$$

If there is no prime p for which $\mathfrak{F}$ has characteristic p, $\mathfrak{F}$ is said to have characteristic 0. For every prime p the field $\mathbb{Z}/(p)$ of the integers modulo p has characteristic p. The field $\mathfrak{R}$ of real numbers has characteristic 0.

$\text{Mod}_{S_{\text{ar}}}(\varphi_F \wedge \chi_p)$ is the *class of fields of characteristic p*. Hence this class is *elementary. The class of fields of characteristic* 0 *is* Δ-*elementary*; it can be represented as $\text{Mod}_{S_{\text{ar}}}(\{\varphi_F\} \cup \{\neg \chi_p \mid p \text{ is prime}\})$. The following consideration will show that it is not elementary. Let φ be an S_{ar}-sentence which is valid in all fields of characteristic 0, that is,

$$\{\varphi_F\} \cup \{\neg \chi_p \mid p \text{ is prime}\} \models \varphi.$$

By the compactness theorem there is an n_0 (depending on φ) such that

$$\{\varphi_F\} \cup \{\neg \chi_p \mid p \text{ is prime}, p < n_0\} \models \varphi.$$

Hence φ is valid in all fields of characteristic $\geq n_0$.

Thus we have proved:

3.3 Theorem. *An S_{ar}-sentence which is valid in all fields of characteristic* 0 *is valid in all fields whose characteristic is sufficiently large.* □

We conclude from this that the class of fields of characteristic 0 is not elementary, for otherwise, there would have to be a sentence φ which is valid precisely in the fields of characteristic 0.

As an instance of 3.3 one obtains the well-known algebraic result that two polynomials $\rho(x)$ and $\sigma(x)$, whose coefficients are integral multiples of the unit element and which are relatively prime over all fields of characteristic 0, are also relatively prime over all fields of sufficiently large characteristic. In order to verify this, one rewrites the statement that $\rho(x)$ and $\sigma(x)$ are relatively prime as an S_{ar}-sentence. In the case $\rho(x) := 3x^2 + 1$ and $\sigma(x) := x^3 - 1$ one can take the sentence

$$\begin{aligned}
&\neg \exists u_0 \exists u_1 \exists w_0 \exists w_1 \exists z_0 \exists z_1 \exists z_2 \forall x \\
&\qquad\quad ((u_0 + u_1 \cdot x) \cdot (w_0 + w_1 \cdot x) \equiv (1 + 1 + 1) \cdot x \cdot x + 1 \\
&\qquad \wedge (u_0 + u_1 \cdot x) \cdot (z_0 + z_1 \cdot x + z_2 \cdot x \cdot x) \equiv x \cdot x \cdot x - 1) \\
&\wedge \neg \exists u_0 \exists u_1 \forall x\, (u_0 + u_1 \cdot x) \cdot ((1 + 1 + 1) \cdot x \cdot x + 1) \equiv x \cdot x \cdot x - 1.
\end{aligned}$$

Here "$\ldots \equiv x \cdot x \cdot x - 1$" stands for "$\ldots + 1 \equiv x \cdot x \cdot x$" (the symbol $-$ does not belong to S_{ar}!).

3.4. *The class of finite S-structures* (*for a fixed S*), *the class of finite groups, and the class of finite fields are not Δ-elementary.* The proof is simple: If, for example, the class of finite fields were of the form $\mathrm{Mod}_{S_{\mathrm{ar}}} \Phi$, then Φ would be a set of sentences having arbitrarily large finite models (e.g., the fields of the form $\mathbb{Z}/(p)$) but no infinite model. That would contradict 2.2. □

On the other hand, exercise 3.6 below shows that the corresponding classes of *infinite* S-structures (groups, fields) are Δ-elementary.

3.5. *The class of torsion groups is not Δ-elementary.* We give an indirect proof, assuming (for a suitable set Φ of S_{gr}-sentences) $\mathrm{Mod}_{S_{\mathrm{gr}}} \Phi$ to be the class of torsion groups. Let

$$\Psi := \Phi \cup \{\neg \underbrace{x \circ \cdots \circ x}_{n\text{-times}} \equiv e \mid n \geq 1\}.$$

Every finite subset Ψ_0 of Ψ has a model: Choose an n_0 such that

$$\Psi_0 \subset \Phi \cup \{\neg \underbrace{x \circ \ldots \circ x}_{n\text{-times}} \equiv e \mid 1 \leq n < n_0\};$$

then every cyclic group of order n_0 is a model of Ψ_0 if x is interpreted by a generating element. Now let $(\mathfrak{G}, \beta)$ be a model of Ψ. Then $\beta(x)$ does not have finite order, showing that $\mathfrak{G}$ is a model of Φ but not a torsion group. □

3.6 Exercise. Let $\mathfrak{K}$ be a Δ-elementary class of structures. Show that the class $\mathfrak{K}^\infty$ of structures in $\mathfrak{K}$ with infinite domain is also Δ-elementary.

3.7 Exercise. If $\mathfrak{K}$ is a class of S-structures, $\Phi \subset L_0^S$ and $\mathfrak{K} = \text{Mod}_S\,\Phi$, then Φ is said to be a *system of axioms* for $\mathfrak{K}$. Show:

(a) $\mathfrak{K}$ is elementary if and only if there is a finite system of axioms for $\mathfrak{K}$.
(b) If $\mathfrak{K}$ is elementary and $\mathfrak{K} = \text{Mod}_S\,\Phi$ then there is a finite subset Φ_0 of Φ such that $\mathfrak{K} = \text{Mod}_S\,\Phi_0$.

3.8 Exercise. A set Φ of S-sentences is called *independent* if no $\varphi \in \Phi$ is a consequence of $\Phi - \{\varphi\}$. Show:

(a) Every finite set Φ of S-sentences has an independent subset Φ_0 such that $\text{Mod}_S\,\Phi = \text{Mod}_S\,\Phi_0$.
(b) If S is at most countable then every Δ-elementary class of S-structures has an independent system of axioms. (*Hint*: Start by defining an axiom system $\varphi_0, \varphi_1, \varphi_2, \ldots$ such that $\models \varphi_{i+1} \rightarrow \varphi_i$ for $i \in \mathbb{N}$.)

3.9 Exercise. Let Φ be the finite system of axioms for vector spaces expressed in terms of $S = \{\underline{F}, \underline{V}, +, \cdot, 0, 1, \oplus, e, *\}$ (cf. III.7.2). Show:

(a) For every n the class of n-dimensional vector spaces is elementary.
(b) The class of infinite-dimensional vector spaces is Δ-elementary.
(c) The class of finite-dimensional vector spaces is not Δ-elementary.

§4. Elementarily Equivalent Structures

We begin by introducing two new concepts.

4.1 Definition. (a) Two S-structures $\mathfrak{A}$ and $\mathfrak{B}$ are called *elementarily equivalent* (written: $\mathfrak{A} \equiv \mathfrak{B}$) if for every S-sentence φ we have $\mathfrak{A} \models \varphi$ iff $\mathfrak{B} \models \varphi$.
(b) For an S-structure $\mathfrak{A}$ let $\text{Th}(\mathfrak{A}) := \{\varphi \in L_0^S \mid \mathfrak{A} \models \varphi\}$. $\text{Th}(\mathfrak{A})$ is called the (first-order) *theory* of $\mathfrak{A}$.

4.2 Lemma. *For S-structures $\mathfrak{A}$ and $\mathfrak{B}$,*

$$\mathfrak{B} \equiv \mathfrak{A} \quad \textit{iff} \quad \mathfrak{B} \models \text{Th}(\mathfrak{A}).$$

Proof. If $\mathfrak{B} \equiv \mathfrak{A}$ then, since $\mathfrak{A} \models \mathrm{Th}(\mathfrak{A})$, also $\mathfrak{B} \models \mathrm{Th}(\mathfrak{A})$. Conversely, if $\mathfrak{B} \models \mathrm{Th}(\mathfrak{A})$ then, given an S-sentence φ, we examine the two possibilities: (i) If $\mathfrak{A} \models \varphi$ then $\varphi \in \mathrm{Th}(\mathfrak{A})$ and hence $\mathfrak{B} \models \varphi$. (ii) If not $\mathfrak{A} \models \varphi$, then $\neg\varphi \in \mathrm{Th}(\mathfrak{A})$; thus $\mathfrak{B} \models \neg\varphi$ and therefore not $\mathfrak{B} \models \varphi$. □

In the following let $\mathfrak{A}$ be a fixed S-structure. We consider

(1) the class $\{\mathfrak{B} \mid \mathfrak{B} \cong \mathfrak{A}\}$ of structures isomorphic to $\mathfrak{A}$;
(2) the class of structures which satisfy the same sentences as $\mathfrak{A}$, i.e., the class $\{\mathfrak{B} \mid \mathfrak{B} \equiv \mathfrak{A}\}$ of structures elementarily equivalent to $\mathfrak{A}$.

From the isomorphism lemma III.5.5 it follows directly that isomorphic structures are elementarily equivalent, that is,

$$(+) \qquad \{\mathfrak{B} \mid \mathfrak{B} \cong \mathfrak{A}\} \subset \{\mathfrak{B} \mid \mathfrak{B} \equiv \mathfrak{A}\}.$$

4.3 Theorem. (a) *If* $\mathfrak{A}$ is infinite *then the class* $\{\mathfrak{B} \mid \mathfrak{B} \cong \mathfrak{A}\}$ *is not* Δ*-elementary; in other words, no infinite structure can be characterized up to isomorphism in a first-order language.*
(b) *The class* $\{\mathfrak{B} \mid \mathfrak{B} \equiv \mathfrak{A}\}$ *is* Δ*-elementary; in fact* $\{\mathfrak{B} \mid \mathfrak{B} \equiv \mathfrak{A}\} = \mathrm{Mod}_S\, \mathrm{Th}(\mathfrak{A})$. *Moreover,* $\{\mathfrak{B} \mid \mathfrak{B} \equiv \mathfrak{A}\}$ *is the smallest* Δ*-elementary class which contains* $\mathfrak{A}$.

From 4.3 together with (+) we obtain the result that for infinite $\mathfrak{A}$ the class $\{\mathfrak{B} \mid \mathfrak{B} \cong \mathfrak{A}\}$ must be a proper subclass of $\{\mathfrak{B} \mid \mathfrak{B} \equiv \mathfrak{A}\}$; in particular:

4.4 Corollary. *For each infinite structure there exists an elementarily equivalent, nonisomorphic structure.* □

Proof of 4.3. (a) We assume $\mathfrak{A}$ to be infinite and Φ to be a set of S-sentences such that

$$(*) \qquad \mathrm{Mod}_S\, \Phi = \{\mathfrak{B} \mid \mathfrak{B} \cong \mathfrak{A}\}.$$

Φ has an infinite model, and hence by 2.3, it has a model $\mathfrak{B}$ with at least as many elements as the power set of $\mathfrak{A}$. Hence $\mathfrak{B}$ is not isomorphic to $\mathfrak{A}$, in contradiction to $(*)$.

(b) From 4.2 it follows immediately that $\{\mathfrak{B} \mid \mathfrak{B} \equiv \mathfrak{A}\} = \mathrm{Mod}_S\, \mathrm{Th}(\mathfrak{A})$. Now, if $\mathrm{Mod}_S\, \Phi$ is another Δ-elementary class containing $\mathfrak{A}$, then $\mathfrak{A} \models \Phi$ and therefore $\mathfrak{B} \models \Phi$ for every $\mathfrak{B}$ such that $\mathfrak{B} \equiv \mathfrak{A}$; hence $\{\mathfrak{B} \mid \mathfrak{B} \equiv \mathfrak{A}\} \subset \mathrm{Mod}_S\, \Phi$. □

4.3(b) shows that a Δ-elementary class contains, together with any given structure all elementarily equivalent ones. In certain cases one can use this fact to show that a class $\mathfrak{K}$ is not Δ-elementary. To do this one simply specifies two elementarily equivalent structures, one of which belongs to $\mathfrak{K}$, and the other does not. We illustrate this method in the case of archimedean fields.

An ordered field $\mathfrak{F}$ is called *archimedean* if for every $a \in F$ there is a natural number n such that

$$a <^F \underbrace{1^F + \cdots + 1^F}_{n\text{-times}}.$$

For example, the ordered field of rational numbers and the ordered field $\mathfrak{R}^<$ of real numbers are archimedean. We show that there is an ordered field elementarily equivalent to $\mathfrak{R}^<$ which is not archimedean. Then we shall have obtained:

4.5 Theorem. *The class of archimedean fields is not Δ-elementary.*

Proof. Let

$$\Psi = \mathrm{Th}(\mathfrak{R}^<) \cup \{\underline{0} < x, \underline{1} < x, \underline{2} < x, \ldots\},$$

where $\underline{0}, \underline{1}, \underline{2}, \ldots$ stand for the S_{ar}-terms $0, 1, 1 + 1, \ldots$. Every finite subset of Ψ is satisfiable, for example, by an interpretation of the form $(\mathfrak{R}^<, \beta)$, where $\beta(x)$ is a sufficiently large natural number. By the compactness theorem there is a model $(\mathfrak{B}, \beta')$ of Ψ. Since $\mathfrak{B} \models \mathrm{Th}(\mathfrak{R}^<)$, $\mathfrak{B}$ is an ordered field elementarily equivalent to $\mathfrak{R}^<$, but (as shown by the element $\beta'(x)$) it is not archimedean. □

The application of the compactness theorem in the preceding proof is typical and has already been used several times (cf. 2.3, 3.5). In each case the problem consists in finding a structure with certain properties which can be expressed in first-order language by means of a suitable set Ψ of formulas. To prove satisfiability of Ψ one employs the compactness theorem. In the preceding proof Ψ contains (in addition to $\mathrm{Th}(\mathfrak{R}^<)$) formulas which guarantee that there is an element which violates the archimedean ordering property. The compactness theorem says in this case that, from the existence of ordered fields with arbitrarily large "finite" elements, one can conclude the existence of an ordered field with an "infinitely large" element. We shall give some further examples.

The axiom system Π from III.7.5 characterizes the structure $\mathfrak{N}$ up to isomorphism. However, $\mathfrak{N}$ cannot be characterized up to isomorphism by means of first-order formulas (cf. 4.4). Hence the induction axiom, being the only second-order axiom of Π, cannot be formulated as a first-order formula or as a set of first-order formulas.

A structure which is elementarily equivalent, but not isomorphic, to $\mathfrak{N}$ is called a *nonstandard model of arithmetic*. The proof of 2.3 shows that there exists an *uncountable* nonstandard model of arithmetic. We now prove:

4.6 Skolem's Theorem. *There is a countable nonstandard model of arithmetic.*

Proof. Let

$$\Psi := \mathrm{Th}(\mathfrak{N}) \cup \{\neg x \equiv \underline{0}, \neg x \equiv \underline{1}, \neg x \equiv \underline{2}, \ldots\}.$$

Every finite subset of Ψ has a model of the form $(\mathfrak{N}, \beta)$, where $\beta(x)$ is a sufficiently large natural number. By the compactness theorem there is a model $(\mathfrak{A}, \beta)$ of Ψ, which by the Löwenheim–Skolem theorem we may assume to be at most countable. $\mathfrak{A}$ is a structure elementarily equivalent to $\mathfrak{N}$. Since for $m \neq n$ the sentence $\neg \underline{m} \equiv \underline{n}$ belongs to $\mathrm{Th}(\mathfrak{N})$, $\mathfrak{A}$ is infinite and hence it is countable. $\mathfrak{N}$ and $\mathfrak{A}$ are not isomorphic, since an isomorphism π from $\mathfrak{N}$ onto $\mathfrak{A}$ would have to map $n = \underline{n}^{\mathfrak{N}}$ to $\underline{n}^{\mathfrak{A}}$ (cf. (i) in the proof of III.5.5), and thus $\beta(x)$ would not lie in the image of π. □

Considering the set $\mathrm{Th}(\mathfrak{N}^<) \cup \{\neg x \equiv \underline{0}, \ \neg x \equiv \underline{1}, \ \neg x \equiv \underline{2}, \ldots\}$, we obtain analogously:

4.7 Theorem. *There is a countable structure elementarily equivalent to $\mathfrak{N}^<$ which is not isomorphic to $\mathfrak{N}^<$. (In other words, there is a countable nonstandard model of* $\mathrm{Th}(\mathfrak{N}^<)$.) □

What do nonstandard models of $\mathrm{Th}(\mathfrak{N})$ or $\mathrm{Th}(\mathfrak{N}^<)$ look like? In the following we gain some insight into the order structure of a nonstandard model $\mathfrak{A}$ of $\mathrm{Th}(\mathfrak{N}^<)$ (and hence also into the structure of a nonstandard model of $\mathrm{Th}(\mathfrak{N})$; cf. exercise 4.9).

In $\mathfrak{N}^<$ the sentences

$$\forall x(\underline{0} \equiv x \vee \underline{0} < x),$$

$$\underline{0} < \underline{1} \wedge \forall x(\underline{0} < x \rightarrow \underline{1} \equiv x \vee \underline{1} < x),$$

$$\underline{1} < \underline{2} \wedge \forall x(\underline{1} < x \rightarrow \underline{2} \equiv x \vee \underline{2} < x), \ldots$$

hold. They say that 0 is the smallest element, 1 the next smallest element after 0, 2 the next smallest element after 1, and so on. Since these sentences also hold in $\mathfrak{A}$, the "initial segment" of $\mathfrak{A}$ looks as follows:

$\underline{0}^A \quad \underline{1}^A \quad \underline{2}^A \quad \underline{3}^A \quad \ldots$

In addition, $\mathfrak{A}$ contains a further element, say a, since otherwise $\mathfrak{A}$ and $\mathfrak{N}$ would be isomorphic. Furthermore, $\mathfrak{N}$ and hence $\mathfrak{A}$ satisfy a sentence φ which says that for every element there is an immediate successor and for every element other than 0 there is an immediate predecessor. From this it follows easily that A contains, in addition to a, infinitely many other elements which together with a are ordered by $<^A$ like the integers:

$\underline{0}^A \quad \underline{1}^A \quad \underline{2}^A \quad \ldots \quad \ldots \quad a \quad \ldots$

If we consider the element $a +^A a$ we are led to further elements of A:

$\underline{0}^A \quad \underline{1}^A \quad \underline{2}^A \quad \ldots \quad a \quad \ldots \quad a +^A a \quad \ldots$

The reader should give a proof of this and also verify that between every two copies of $(\mathbb{Z}, <^{\mathbb{Z}})$ there lies another copy.

The examples in this and the previous section show that there are important classes of structures which cannot be axiomatized in a first-order language. On the other hand, this weakness of expressive power also has agreeable consequences. For example, the argument establishing that the class of archimedean fields is not axiomatizable yields a proof of the existence of non-archimedean ordered fields; and the fact that the class of fields of characteristic 0 cannot be axiomatized by means of a single S_{ar}-sentence is complemented by the interesting result 3.3. Using similar methods, one can obtain structures elementarily equivalent to the ordered field $\mathfrak{R}^<$ of real numbers which contain, in addition to the real numbers, infinitely large elements and infinitely small positive elements (infinitesimals). Such structures can be used in a development of analysis which avoids the $\varepsilon - \delta$-technique (*nonstandard analysis*; cf. [14], [23]). Thus we see that the first-order languages turn out to be a useful tool in various areas of mathematics. Semantic investigations of this kind belong to the subject called *model theory*. We refer the reader to [4] for further information.

4.8 Exercise. Show that a sentence which is valid in all non-archimedean ordered fields is valid in all ordered fields.

4.9 Exercise. Let the S_{ar}-structure $\mathfrak{A}$ be a model of $\mathrm{Th}(\mathfrak{N})$. Let the binary relation $<^A$ be defined as follows:

For all $a, b \in A$, $a <^A b$ iff $a \neq b$ and there is $c \in A$ such that $a +^A c = b$.

Show that $(\mathfrak{A}, <^A)$ is a model of $\mathrm{Th}(\mathfrak{N}^<)$.

4.10 Exercise. If $\mathfrak{A}$ is a model of arithmetic (that is, $\mathfrak{A} \models \mathrm{Th}(\mathfrak{N})$) and if $a, b \in A$, then a is said to be a divisor of b (written: $a|b$) if $a \cdot^A c = b$ for a suitable $c \in A$. Let Q be a set of prime numbers. Show that there is a model $\mathfrak{A}$ of arithmetic which contains an element a whose prime divisors are just the members of Q, that is, for every prime p:

$$\underbrace{1^A + \cdots + 1^A}_{p\text{-times}} | a \quad \text{iff } p \in Q.$$

Conclude that there are uncountably many pairwise nonisomorphic countable models of arithmetic.

4.11 Exercise. Let $\mathfrak{A} = (A, <^A)$ be a partially defined ordering (cf. III.6.4). We say that $<^A$ (or also $(A, <^A)$) has an *infinite descending chain*, if there are elements $a_0, a_1, a_2, \ldots$ in the field of $<^A$ such that $\ldots <^A a_2 <^A a_1 <^A a_0$.

Show:

(a) $(\mathbb{N}, <^{\mathbb{N}})$ contains no infinite descending chain; on the other hand, if A is a nonstandard model of $\mathrm{Th}(\mathfrak{N}^{<})$, then $(A, <^{A})$ contains an infinite descending chain.
(b) Let $< \in S$ and $\Phi \subset L_0^S$. Assume that for every $m \in \mathbb{N}$ there is a model $\mathfrak{A}$ of Φ such that $(A, <^{A})$ is a partially defined ordering and the field of $<^{A}$ contains at least m elements. Then there exists also a model $\mathfrak{B}$ of Φ such that $(B, <^{B})$ is a partially defined ordering containing an infinite descending chain.

CHAPTER VII

The Scope of First-Order Logic

In the introductory chapter we realized that investigations into the logical reasoning used in mathematics require an analysis of the concepts of mathematical proposition and proof. In undertaking such an analysis, we were led to introduce the first-order languages. Further we defined a notion of formal proof which corresponds to the intuitive concept of mathematical proof. The completeness theorem then shows that every proposition which is mathematically provable from a system of axioms (and thus follows from it) can also be obtained by means of a formal proof, provided the proposition and the system of axioms admit a first-order formulation.

In this chapter we discuss what has been achieved so far and what implications this has for the foundations of mathematics. To start our discussion let us consider the following questions:

(1) One goal of our investigations was a clarification of the notion of proof. However, we carried out mathematical proofs before the notion of proof was made precise. Are we not trapped in a vicious circle? Further, even if there are no problems of this kind in our approach, how can we then justify the rules of the sequent calculus $\mathfrak{S}$?

(2) We realized, particularly in Chapter VI, that the first-order languages have certain deficiencies in expressive power. Hence the question: What effect does the restriction to first-order languages have on the scope of our investigations?

We deal with the second question in §2. There we shall see that the first-order languages are in principle sufficient for the mathematics of today. Hence the following discussion, pertaining to the first question, applies, in fact, to the whole of mathematics.

§1. The Notion of Formal Proof

In answering question (1), we want to show that no mathematical proofs are needed to introduce the notion of formal proof. In our discussion we also investigate the nature of the sequent rules and consider possible means of justifying them.

In §2 we shall argue that a finite set S of concretely chosen symbols suffices to represent the statements and arguments arising in mathematics. Therefore in this discussion we can specify the symbols as concrete signs; thus terms, formulas, and sequents are concrete strings of symbols and not abstract mathematical entities such as are, for example, formulas in a language whose symbol set is $\{c_r | r \in \mathbb{R}\}$ (cf. II.1).

The notion of formal proof is based on the manipulation of symbol strings such as terms, formulas, and sequents. These manipulations are governed by a series of calculi, like the calculus of terms and the sequent calculus. The application of rules in these calculi consists of simple syntactic operations. We illustrate this in the case of the sequent calculus. To clarify the aspect we have in mind let us start by a comparison with the rules of chess.

The rules of chess permit certain operations on concrete objects, the chess pieces. Applying a rule, that is, making a move, consists of proceeding from one configuration of pieces to another. Each individual rule of chess is so simple that everyone who knows the rules—even if he is not a chess player—can carry out moves by himself, or can check moves to determine whether they were made according to the rules.

A similar situation pertains in the case of sequent rules. Clearly the rules are motivated by the intended meanings (of $\neg$, $\vee$, $\equiv, \ldots$), but their application does not require any knowledge of these meanings: one merely performs concrete syntactic operations on strings of symbols. Anyone who knows the rules—even if he is not a logician or a mathematician—can apply them and can check whether an application has been carried out correctly. Admittedly, when dealing with sequents, we have often relied on mathematical propositions (for example, we invoked the unique decomposition of a sequent into formulas when speaking of *the* succedent). But this can be avoided if, when applying a rule, we not only note the sequent, but also keep a record of how the symbol strings in it were obtained. We give some examples.

(a) Let Θ_1 and Θ_2 be sequents which occur in a derivation. One reads from the record accompanying the derivation that Θ_1 was obtained by forming a string from $\varphi_0, \ldots, \varphi_m$ and that Θ_2 was obtained similarly from $\psi_0, \ldots, \psi_n$. If one wants to apply the rule ($\vee$A), for example, one must first check whether $n = m \geq 1$, and whether the symbol strings φ_i and ψ_i agree for every $i \neq n - 1$. If so, one can apply ($\vee$A) by forming the symbol string $\varphi_0 \ldots \varphi_{n-2}(\varphi_{n-1} \vee \psi_{n-1})\varphi_n$ from the components $\varphi_0, \ldots, \varphi_{n-2}, \varphi_{n-1}, \psi_{n-1}, \varphi_n$, (, $\vee$, and). Moreover, one notes in the record that this symbol string was obtained from the components $\varphi_0, \ldots, \varphi_{n-2}, (\varphi_{n-1} \vee \psi_{n-1})$, and φ_n.

(b) An application of the rule ($\equiv$) consists of writing down a sequent of the form $t \equiv t$, where the term t, for its part, has to be given by means of a derivation in the calculus of terms (cf. II.3.1).

(c) Similarly, when one uses the rule ($\exists$A) to proceed from the sequent $\Gamma\ \varphi\frac{y}{x}\ \psi$ to the sequent $\Gamma\ \exists x\varphi\ \psi$, one must supply a derivation of $\varphi x y \varphi\frac{y}{x}$ in the substitution calculus (cf. III.8.11), and, for every χ in $\Gamma\ \exists x\varphi\ \psi$, one must supply a derivation of $y\ \chi$ in the calculus of nonfree occurrence for variables (cf. II.5.2) in order to show that the condition "y is not free in $\Gamma\ \exists x\varphi\ \psi$" is fulfilled. Then, starting from the sequent $\Gamma\ \varphi\frac{y}{x}\ \psi$, one needs only to write down the sequent $\Gamma\ \exists x\varphi\ \psi$.

From these examples it becomes clear that an application of the sequent rules consists of purely syntactic manipulations which can be carried out without any reference to mathematical arguments. Since, by definition, a formal proof is just a sequence consisting of sequents, each of which is obtained by an application of a sequent rule to preceding sequents, it is obvious from our previous remarks that no mathematical proofs are needed in order to introduce the notion of formal proof. Thus our approach is not circular. The proofs we have given before defining the notion of formal proof, and the mathematical arguments we have used in building up the semantics, merely served the purpose of gaining insight into first-order languages and of motivating our development.

However, a word of warning is in order when considering this reduction of the notion of proof to a triviality by the calculus of sequents: We have seen that no mathematical talent, only patience, is needed to verify a formal proof in accordance with the rules; but it is a completely different matter to understand the idea of a proof, not to speak of developing such ideas oneself. Likewise, in chess there is also a great difference between knowing the rules and being able to checkmate a skillful opponent. Thus when determining the notion of formal proof we did not really touch upon the more creative part of mathematical activity (and this includes not only the development of proof ideas, but also the introduction of adequate concepts, setting up suitable systems of axioms, and finding new interesting conjectures).

Does our formal notion of proof provide a *justification* of common mathematical reasoning? Certainly not; for we have merely imitated methods of proof in the framework of a precisely defined language. However, we can at least claim that the sequent rules correspond to the normal usage of connectives, quantifiers, and equality in mathematics. For example, the $\vee$-rules reflect the use of the inclusive "or", according to which the disjunction of two propositions is true if and only if at least one of the propositions is true. Admittedly, such usage of "or" rests on certain assumptions; for example, it must be meaningful to speak of the truth or falsehood of a mathematical proposition, and every such proposition must be either true or false (tertium non datur). In traditional mathematics (which in this regard is also called *classical* mathematics) these assumptions are accepted. Thus the rules of the sequent calculus are based upon the classical usage of the logical connectives.

Some mathematicians engaged in foundational questions, among them *intuitionists*, do not share the classical point of view. An intuitionist associates with the assertion of a mathematical proposition the requirement that it be proved in a "constructive" way. For instance, an existential statement must be proved by presenting an example, and a disjunction must be proved by establishing one of its members. To illustrate this we consider the following two statements.

> A: Every even number ≥ 4 is the sum of two primes (Goldbach's conjecture);
>
> not A: Not every even number ≥ 4 is the sum of two primes.

From the classical point of view (A or not A) is true. However, an intuitionist cannot assert (A or not A) since neither the proposition A nor the proposition (not A) has hitherto been proved (even using classical methods).

This example already shows that mathematics as pursued by an intuitionist, the so-called *intuitionistic mathematics* (cf. [17]), differs considerably from classical mathematics. Intuitionists investigate "mental mathematical constructions as such, without reference to questions regarding the nature of the constructed objects, such as whether these objects exist independently of our knowledge of them" (cf. [17], p. 1). By contrast, some mathematicians adopt the classical point of view from the conviction that "the objects in mathematics, together with the mathematical domains, exist as such, like the platonic ideas" ([24], p. 1), i.e., that propositions concerning these objects describe properties which either do or do not hold, and hence are either true or false.

We see from this discussion that the possibilities for justifying methods of mathematical reasoning (and specifically for justifying a proof calculus) depend essentially on epistemological assumptions. We shall continue to adopt the classical point of view.

§2. Mathematics Within the Framework of First-Order Logic

In this section we wish to discuss the latter question raised at the beginning of the chapter: How serious is the restriction to first-order languages?

To treat this question we start with the example of arithmetic. In this case, the weakness of the expressive power of first-order languages manifests itself in the fact that the structure $\mathfrak{N}_\sigma = (\mathbb{N}, \sigma, 0)$ (cf. III.7.3) cannot be characterized up to isomorphism in $L^{\{\underline{\sigma}, 0\}}$. On the other hand, according to Dedekind's theorem, $\mathfrak{N}_\sigma$ can be characterized in a second-order language by the Peano axioms (cf. III.7.4):

(P1) $\forall x \neg \underline{\sigma} x \equiv 0$.
(P2) $\forall x \forall y(\underline{\sigma} x \equiv \underline{\sigma} y \rightarrow x \equiv y)$.
(P3) $\forall X((X0 \wedge \forall x(Xx \rightarrow X\underline{\sigma} x)) \rightarrow \forall y\, Xy)$.

Let us call a structure which satisfies (P1)–(P3) a *Peano structure*. Then we can formulate Dedekind's theorem as follows:

2.1. *Any two Peano structures are isomorphic.*

Since Peano structures cannot be characterized in the first-order language, one might suspect that the result 2.1 cannot be formulated in the framework given by first-order logic, and in particular, that its proof in III.7.4, which involves (P1)–(P3), cannot be carried out within this framework. Nevertheless this can be achieved as we now show.

First let us note that in 2.1 a statement is made about $\{\underline{\sigma}, 0\}$-structures. We want to interpret 2.1 as a statement about a domain which comprises as elements all Peano structures and also with any two such structures an isomorphism between them. Furthermore this domain should contain the elements and subsets of Peano structures, since these also play a rôle in the formulation of (P1)–(P3) and in the proof of 2.1.

In order to avoid drawing arbitrary boundaries and to enable us to apply our discussion to other propositions besides 2.1, we shall consider as domain the totality of *all* objects which are treated in mathematics; this we shall call the (mathematical) *universe*. The universe contains not only "simple" objects, such as the natural numbers or the points of the euclidean plane, but also "more complicated" objects, such as sets, functions, structures, or topological spaces. A mathematician assumes in his arguments that this universe has certain properties: for example, that for every two objects a_1 and a_2 the set $\{a_1, a_2\}$ exists, likewise for any two sets M_1, M_2 the union $M_1 \cup M_2$, and for every injective function f the inverse f^{-1}. Mathematical statements can then be regarded as propositions about the universe. From this point of view, 2.1 says that for every two Peano structures $\mathfrak{A}$ and $\mathfrak{B}$ in the universe there is another object in the universe which is an isomorphism between $\mathfrak{A}$ and $\mathfrak{B}$.

Now it is possible to present in a suitable first-order language a rather simple set of sentences expressing all the properties of the universe which mathematicians use. Proposition 2.1 can also be formalized in this language. In other words, 2.1 can be formalized as a proposition about the universe in a first-order language L^S appropriate to the universe, just as the proposition "there is no largest real number" can be formalized as a proposition about the structure $(\mathbb{R}, <^{\mathbb{R}})$ in the language $L^{\{<\}}$ appropriate to $(\mathbb{R}, <^{\mathbb{R}})$.

We carry out some steps of this idea more carefully: A preliminary analysis of the totality of mathematical objects leads us to a symbol set which is suitable for the universe. In a second step we present parts of a system Φ_0 of axioms which express those properties of the universe used in mathematics. (A complete presentation of such a system Φ_0 follows in §3.) Finally, we indicate how to obtain a first-order formalization of 2.1.

When introducing the universe, we spoke of "simple" objects (numbers, points, ...) and "complex" objects (sets, functions, ...). For the sake of simplicity we make use of the empirical fact that the whole spectrum of "complex" objects can be reduced to the concept of set. (We shall carry out this reduction for ordered pairs and functions.) We call the "simple" objects *urelements*. Thus, the universe contains only urelements and sets. The sets consist of elements which are either urelements or else sets themselves. Therefore Φ_0 essentially collects basic properties of sets and hence is called a *system of axioms for set theory*.

We use the unary relation symbols $\underline{U}$ ("... is an urelement") and $\underline{M}$ ("... is a set") to distinguish between urelements and sets, and we use the binary relation symbol $\underline{\in}$ for the relation "is an element of..." . Thus we are led to the symbol set $S := \{\underline{U}, \underline{M}, \underline{\in}\}$.

Now we give four axioms from Φ_0 which formalize simple properties of the universe.

(A0) $\forall x(\underline{U}x \vee \underline{M}x)$
"Every object is an urelement or a set".

(A1) $\forall x \neg(\underline{U}x \wedge \underline{M}x)$
"No object is both an urelement and a set".

(A2) $\forall x \forall y((\underline{M}x \wedge \underline{M}y \wedge \forall z(z \underline{\in} x \leftrightarrow z \underline{\in} y)) \rightarrow x \equiv y)$
"Two sets which contain the same elements are equal".

(A3) $\forall x \forall y \exists z(\underline{M}z \wedge \forall u(u \underline{\in} z \leftrightarrow (u \equiv x \vee u \equiv y)))$
"For every two objects x and y, the pair set $\{x, y\}$ exists".

The set z, whose existence is guaranteed by (A3), is uniquely determined by (A2). Repeated application of (A3) yields the existence of the set $\{\{x, x\}, \{x, y\}\}$. This set is normally written (x, y) and called the *ordered pair* of x and y. It is not difficult to show from (A0)–(A3) that

$$(x, y) = (x', y') \quad \text{iff } x = x' \quad \text{and} \quad y = y'.$$

Ordered triples can then be introduced by

$$(x, y, z) := ((x, y), z).$$

In order to obtain formalizations which are easier to read, we introduce a number of abbreviations.

($\subseteq$) $x \subseteq y$ for $\underline{M}x \wedge \underline{M}y \wedge \forall z(z \underline{\in} x \to z \underline{\in} y)$.
("x is a subset of y")

(Instead of treating "$x \subseteq y$" as an abbreviation for a formula of L^S we could have added the binary relation symbol $\subseteq$ to S and expanded Φ_0 by adding the axiom

$$\forall x\, \forall y(x \subseteq y \leftrightarrow (\underline{M}x \wedge \underline{M}y \wedge \forall z(z \underline{\in} x \to z \underline{\in} y))).$$

Both approaches are equivalent, as we shall see in VIII.1.)

($\underline{OP}$) $\underline{OP}zxy$ for $\underline{M}z \wedge \forall u(u \underline{\in} z \leftrightarrow (\underline{M}u \wedge (\forall v(v \underline{\in} u \leftrightarrow v \equiv x) \vee \forall v(v \underline{\in} u \leftrightarrow (v \equiv x \vee v \equiv y)))))$
("z is the ordered pair of x and y")

($\underline{OT}$) $\underline{OT}uxyz$ for $\underline{M}u \wedge \exists v(\underline{OP}uvz \wedge \underline{OP}vxy)$
("u is the ordered triple (x, y, z) as defined above")

($\underline{E}$) $\underline{E}uxy$ for $\underline{M}u \wedge \exists z(z \underline{\in} u \wedge \underline{OP}zxy)$
("The ordered pair (x, y) is an element of u")

($\underline{F}$) $\underline{F}u$ for $\underline{M}u \wedge \forall z(z \underline{\in} u \to \exists x\, \exists y\, \underline{OP}zxy) \wedge \forall x\, \forall y\, \forall y'((\underline{E}uxy \wedge \underline{E}uxy') \to y \equiv y')$
("u is a function, that is, a set of ordered pairs (x, y), where y is the value of u at x").

By means of ($\underline{F}$) the concept of function is reduced in the usual manner to that of set: a function $f: A \to B$ is considered as the set $\{(x, f(x)) \mid x \in A\}$, which is also referred to as the *graph* of f.

($\underline{D}$) $\underline{D}uv$ for $\underline{F}u \wedge \underline{M}v \wedge \forall x(x \underline{\in} v \leftrightarrow \exists y\, \underline{E}uxy)$
("v is the domain of the function u")

($\underline{R}$) $\underline{R}uv$ for $\underline{F}u \wedge \underline{M}v \wedge \forall y(y \underline{\in} v \leftrightarrow \exists x\, \underline{E}uxy)$
("v is the range of the function u").

For simplicity we regard a $\{\underline{\sigma}, 0\}$-structure as an ordered triple (x, y, z) consisting of a set x, a function y: $x \to x$, and an element z of x. Then the following abbreviation "$\underline{PS}u$" expresses that u is a Peano structure, whereby parts (1), (2), and (3) are formulations of the Peano axioms (P1), (P2), and (P3), respectively.

($\underline{PS}$) $\underline{PS}u$ for $\exists x\, \exists y\, \exists z(\underline{OT}uxyz \wedge \underline{M}x \wedge z \underline{\in} x$
$\wedge\, \underline{F}y \wedge \underline{D}yx \wedge \exists v(\underline{R}yv \wedge v \subseteq x) \wedge$
(1) $\forall w\, \neg \underline{E}ywz \wedge$
(2) $\forall w\, \forall w'\, \forall v((\underline{E}ywv \wedge \underline{E}yw'v) \to w \equiv w') \wedge$
(3) $\forall x'((x' \subseteq x \wedge z \underline{\in} x'$
$\wedge\, \forall w\, \forall v((w \underline{\in} x' \wedge \underline{E}ywv) \to v \underline{\in} x')) \to x' \equiv x))$.

The final abbreviation "$\underline{I}wuu'$" states the property that w is an isomorphism of the Peano structure u onto the Peano structure u':

$(\underline{I})$ $\underline{I}wuu'$ for $\underline{PS}u \wedge \underline{PS}u' \wedge \underline{F}w$

$$\wedge \exists x \exists y \exists z \exists x' \exists y' \exists z'(\underline{OT}uxyz \wedge \underline{OT}u'x'y'z'$$
$$\wedge \underline{D}wx \wedge \underline{R}wx'$$
$$\wedge \forall r \forall s \forall v'((\underline{E}wrv' \wedge \underline{E}wsv') \rightarrow r \equiv s)$$
$$\wedge \underline{E}wzz' \wedge \forall v \forall v' \forall r((\underline{E}yvr \wedge \underline{E}wvv') \rightarrow \exists r'(\underline{E}wrr' \wedge \underline{E}y'v'r'))).$$

Thus the following is a formalization of 2.1:

$(+)$ $$\forall u \forall v(\underline{PS}u \wedge \underline{PS}v \rightarrow \exists w \underline{I}wuv).$$

Clearly, $(+)$ is a $\{\underline{U}, \underline{M}, \underline{\in}\}$-sentence. So we have attained our goal of formulating 2.1 within a first-order language. This was possible because we did not distinguish between different types of mathematical objects, such as natural numbers and sets of natural numbers, but simply treated all objects in the universe as first-order ones (compare (P3) and (3) in $(\underline{PS})$). We can achieve even more: Recall that the system Φ_0 (which we have given only in part) captures all properties of the universe needed for mathematical reasoning. By rewriting in L^S the proof of Dedekind's theorem 2.1 (cf. III.7), one can obtain a proof that leads from axioms of Φ_0 to the assertion $(+)$ using only sequent rules. Hence we have:

2.2. $\Phi_0 \vdash \forall u \forall v(\underline{PS}u \wedge \underline{PS}v \rightarrow \exists w \underline{I}wuv)$.

This procedure can be generalized:

Experience shows that all mathematical propositions can be formalized in L^S (or in variants of it), *and that mathematically provable propositions have formalizations which are derivable from* Φ_0. *Thus it is in principle possible to imitate all mathematical reasoning in* L^S *using the rules of the sequent calculus. In this sense, first-order logic is sufficient for mathematics.* At the same time this experience shows that the properties of the universe which are expressed in Φ_0 are a sufficient basis for a set-theoretic development of mathematics. Thus Φ_0 is a formalization of the set-theoretic assumptions about the universe upon which the mathematician ultimately relies. Since these set-theoretic assumptions can be viewed as the background for all mathematical considerations, we call Φ_0, in this connection, a system of axioms for *background set theory*.

On the other hand, Φ_0 itself, like any other system of axioms, can also be the object of mathematical investigations. For example, one can ask whether Φ_0 is consistent or study the models of Φ_0. Such a model has the form $\mathfrak{A} = (A, \underline{U}^A, \underline{M}^A, \underline{\in}^A)$ and is, like every structure, an object of the universe, that is, an object in the sense of background set theory. The same is true of the domain A. Thus as an object of the universe, A is distinct from the universe. (In particular the universe is not the domain of a model of Φ_0.)

Nevertheless, in a model $\mathfrak{A} = (A, \underline{U}^A, \underline{M}^A, \underline{\in}^A)$ of Φ_0, all set-theoretical statements hold which are derivable from Φ_0; but note that, for example, $a \underline{\in}^A b$ (for $a, b \in A$) does not mean that a is an element of b, i.e., that $a \in b$ holds.

Let us emphasize once again that Φ_0 plays two rôles: It is both an object of mathematical investigations and a formalized description of basic properties of the universe. In other words, it is both a mathematical object and a framework for mathematics.

Thus we have two levels, "object set theory" and "background set theory", which must be carefully distinguished. Many paradoxes arise from a confusion of these two levels. In §4 we shall discuss this in more detail. For the present we merely mention *Skolem's paradox*. It is well known that there are uncountably many sets (for example, there are uncountably many subsets of $\mathbb{N}$). This fact can be formalized by a sentence φ, which is derivable from Φ_0. By the Löwenheim–Skolem theorem there is a countable model $\mathfrak{A}$ of Φ_0 and hence of φ. The *countable* model $\mathfrak{A}$ thus satisfies a sentence which says that there are *uncountably* many sets in $\mathfrak{A}$!

§3. The Zermelo–Fraenkel Axioms for Set Theory

We now present in full a system of axioms for set theory. For a more detailed exposition we refer the reader to [8] or [9].

In §2 we assumed that the universe consists only of sets and urelements, and we saw with the help of set-theoretic definitions for concepts such as "ordered pair" and "function" that this assumption is really no restriction. Furthermore, experience has shown that one can even replace the urelements arising in mathematics by suitable sets. Hence in what follows we shall assume that the universe consists only of sets. Later, as an example, we shall give a set-theoretic substitute for the natural numbers.

Since we are abandoning the use of urelements, the symbols $\underline{U}$ and $\underline{M}$ become superfluous. Therefore we formulate the axioms in $L^{\{\underline{\in}\}}$, where the variables are intended to range over the sets of the universe. The resulting system of axioms, called ZFC, is originally due to Zermelo and Fraenkel, and includes the axiom of choice.

ZFC contains the axioms EXT (*the axiom of extensionality*), PAIR (*the pair set axiom*), SUM (*the sum set axiom*), POW (*the power set axiom*), INF (*the axiom of infinity*), AC (*the axiom of choice*), and the axiom schemes SEP (*separation axioms*) and REP (*replacement axioms*).[1]

EXT: $\forall x\, \forall y (\forall z (z \underline{\in} x \leftrightarrow z \underline{\in} y) \to x \equiv y)$.
(Two sets which contain the same elements are equal.)

[1] Often one also includes the so-called axiom of regularity.

SEP: For each $\varphi(z, x_0, \ldots, x_{n-1})$[2] and arbitrary distinct variables x, y which are also distinct from z and the x_i, the axiom

$$\forall x_0 \ldots \forall x_{n-1} \forall x \exists y \forall z(z \underline{\in} y \leftrightarrow (z \underline{\in} x \wedge \varphi(z, x_0, \ldots, x_{n-1}))).$$

(Given a set x and a property P which can be formulated by an $\{\underline{\in}\}$-formula, the set $\{z \in x \mid z$ has the property $P\}$ exists.)

PAIR: $\forall x \forall y \exists z \forall w(w \underline{\in} z \leftrightarrow (w \equiv x \vee w \equiv y))$.
(Given two sets x, y, the pair set $\{x, y\}$ exists.)

SUM: $\forall x \exists y \forall z(z \underline{\in} y \leftrightarrow \exists w(w \underline{\in} x \wedge z \underline{\in} w))$.
(Given a set x, the union of all sets in x exists.)

POW: $\forall x \exists y \forall z(z \underline{\in} y \leftrightarrow \forall w(w \underline{\in} z \rightarrow w \underline{\in} x))$.
(Given a set x, the power set of x exists.)

In order to formulate the remaining axioms more conveniently, we introduce some defined symbols. The considerations in VIII.1 show that formulas which contain these symbols can be regarded as abbreviations of $\{\underline{\in}\}$-formulas. The symbols and their definitions are:

$\underline{\emptyset}$ (constant for the empty set):

$$\forall y(\underline{\emptyset} \equiv y \leftrightarrow \forall z \neg z \underline{\in} y).$$

$\underline{\subseteq}$ (relation symbol for the subset relation):

$$\forall x \forall y(x \underline{\subseteq} y \leftrightarrow \forall z(z \underline{\in} x \rightarrow z \underline{\in} y)).$$

$\underline{\{}, \underline{\}}$ (function symbol for pairing):

$$\forall x \forall y \forall z(\underline{\{}x, y\underline{\}} \equiv z \leftrightarrow \forall w(w \underline{\in} z \leftrightarrow (w \equiv x \vee w \equiv y))).$$

(For the term $\underline{\{}y, y\underline{\}}$ we often write the shorter form $\underline{\{}y\underline{\}}$.)
$\underline{\cup}$ (function symbol for the union):

$$\forall x \forall y \forall z(x \underline{\cup} y \equiv z \leftrightarrow \forall w(w \underline{\in} z \leftrightarrow (w \underline{\in} x \vee w \underline{\in} y))).$$

$\underline{\cap}$ (function symbol for the intersection):

$$\forall x \forall y \forall z(x \underline{\cap} y \equiv z \leftrightarrow \forall w(w \underline{\in} z \leftrightarrow (w \underline{\in} x \wedge w \underline{\in} y))).$$

$\underline{P}$ (function symbol for the power set operation):

$$\forall x \forall y(\underline{P}x \equiv y \leftrightarrow \forall z(z \underline{\in} y \leftrightarrow z \underline{\subseteq} x)).$$

The remaining axioms of ZFC are as follows:

INF: $\exists x(\underline{\emptyset} \underline{\in} x \wedge \forall y(y \underline{\in} x \rightarrow y \underline{\cup} \underline{\{}y\underline{\}} \underline{\in} x))$.
(There exists an infinite set, namely a set containing $\emptyset$, $\{\emptyset\}$, $\{\emptyset, \{\emptyset\}\}, \ldots$.)[3]

[2] Here and in the following we write $\psi(y_0, \ldots, y_{n-1})$ to indicate that the variables occurring free in ψ are among the distinct variables $y_0, \ldots, y_{n-1}$.

[3] At a first glance it might be more natural to demand that there exists a set containing $\emptyset$, $\{\emptyset\}$, $\{\{\emptyset\}\}, \ldots$. This is Zermelo's original version. However, our formulation of INF (due to von Neumann) has become customary because of numerous advantages.

REP: For each $\varphi(x, y, x_0, \ldots, x_{n-1})$ in $L^{\{\in\}}$ and all distinct variables u, v which are also distinct from $x, y, x_0, \ldots, x_{n-1}$ the axiom

$$\forall x_0 \ldots \forall x_{n-1}(\forall x\, \exists^{=1} y\varphi(x, y, x_0, \ldots, x_{n-1}) \\ \rightarrow \forall u\, \exists v\, \forall y(y \,\underset{\sim}{\in}\, v \leftrightarrow \exists x(x \,\underset{\sim}{\in}\, u \wedge \varphi(x, y, x_0, \ldots, x_{n-1})))).$$

(If for parameters $x_0, \ldots, x_{n-1}$ the formula $\varphi(x, y, x_0, \ldots, x_{n-1})$ defines a map $x \mapsto y$, then the image of a set is again a set.)

AC: $$\forall x((\neg \underset{\sim}{\varnothing} \,\underset{\sim}{\in}\, x \wedge \forall u\, \forall v((u \,\underset{\sim}{\in}\, x \wedge v \,\underset{\sim}{\in}\, x \wedge \neg u \equiv v) \rightarrow u \,\underset{\sim}{\cap}\, v \equiv \underset{\sim}{\varnothing})) \\ \rightarrow \exists y\, \forall w(w \,\underset{\sim}{\in}\, x \rightarrow \exists^{=1} z z \,\underset{\sim}{\in}\, w \,\underset{\sim}{\cap}\, y)).$$

(Given a set x of nonempty pairwise disjoint sets, there exists a set which contains exactly one element of each set in x.)

Within the framework of ZFC one can now introduce the notions of ordered pair, ordered triple, function, etc. as we did in the preceding section. Moreover, as already mentioned above, experience shows that ZFC also permits one to substitute suitable sets for urelements, as we demonstrate below in the case of the natural numbers. Thus the insight stated (for Φ_0) in the previous section also applies to ZFC: All mathematical propositions can be formalized in $L^{\{\in\}}$, and provable propositions correspond to sentences derivable from ZFC.

We now effect a set-theoretical substitute for the natural numbers. Moreover in our present framework we exhibit a Peano structure which can play the rôle of $\mathfrak{N}_\sigma$.

The sets $\underset{\sim}{0} := \varnothing$, $\underset{\sim}{1} := \{\varnothing\}$, $\underset{\sim}{2} := \{\varnothing, \{\varnothing\}\}, \ldots$ will play the rôle of the natural numbers 0, 1, 2, ... Thus $\underset{\sim}{0} = \varnothing$, $\underset{\sim}{1} = \{\underset{\sim}{0}\}$, $\underset{\sim}{2} = \{\underset{\sim}{0}, \underset{\sim}{1}\}$, and in general $\underset{\sim}{n} = \{\underset{\sim}{0}, \underset{\sim}{1}, \ldots, \underbrace{n-1}\}$. Let us call a set *inductive* if it contains $\varnothing$, and if whenever it contains x it also contains $x \cup \{x\}$; then the smallest inductive set assumes the rôle of $\mathfrak{N}$. It remains to show that the statement "there is a smallest inductive set" is derivable in ZFC. We give a guideline as to how to proceed. By INF there exists an inductive set, say x. Using SEP we obtain the set

$$\omega := \{z \mid z \in x \text{ and for all inductive } y, z \in y\},$$

which can be shown to be the smallest inductive set. The function $v\colon \omega \rightarrow \omega$, where $v(x) = x \cup \{x\}$ for $x \in \omega$ (i.e., the function $v = \{(x, x \cup \{x\}) \mid x \in \omega\}$) plays the rôle of the successor function. One can then see that $(\omega, v, \underset{\sim}{0})$ is a Peano structure.

We close our presentation of ZFC with an important methodological aspect by briefly discussing the so-called *continuum hypothesis*. This hypothesis was stated at the end of the nineteenth century by G. Cantor and has had a crucial influence on the development of set theory.

Two sets x, y are said to be *of the same cardinality* (written: $x \sim y$) if there is a bijection from x to y. A set is *finite* if and only if it is of the same cardinality as an element of ω; it is *countable* if it is of the same cardinality as ω. The set $\mathbb{R}$ of real numbers (the "continuum") is uncountable (cf. exercise II.1.3).

Now the continuum hypothesis states: Every infinite subset of $\mathbb{R}$ is either countable or of the same cardinality as $\mathbb{R}$.

Using canonically defined symbols $\underline{\mathbb{R}}$, $\underline{\text{Fin}}, \ldots$, this statement can be formulated in $L^{\{\in\}}$ in the following form:

$$\forall x((x \subseteq \underline{\mathbb{R}} \wedge \neg \underline{\text{Fin}}\, x) \rightarrow (\underline{\text{Count}}\, x \vee x \sim \underline{\mathbb{R}})).$$

This formula is denoted by "CH" (Continuum Hypothesis). The question whether the continuum hypothesis holds corresponds to the question whether CH is derivable from ZFC.

K. Gödel showed in 1938:

3.1. *If* ZFC *is consistent then not* ZFC $\vdash \neg$CH,

and P. Cohen showed in 1963:

3.2. *If* ZFC *is consistent then not* ZFC $\vdash$ CH.

Thus if we assume that ZFC is consistent (cf. §4), then neither CH nor $\neg$CH is derivable from it.

According to previous remarks, ZFC embodies our knowledge of the intuitive concept of set which mathematicians in fact use. In the light of the results of Gödel and Cohen we see that our concept is so vague that it does not definitely decide the truth or falsehood of the continuum hypothesis. One can even show (cf. X.7) that it is not possible to give "explicitly" an axiom system Ψ for set theory, which decides every set-theoretic statement ψ (in the sense that either $\Psi \vdash \psi$ or $\Psi \vdash \neg\psi$).

§4. Set Theory as a Basis for Mathematics

In this section we supplement our previous discussion by treating three aspects: In 4.1, taking ZFC as an example, we show how the question of the consistency of mathematics may be made precise by the use of suitable first-order axioms sufficient for mathematics. In 4.2 we discuss misunderstandings which may arise from a confusion of object set theory with background set theory. Finally, in 4.3 we show how first-order logic, like every other mathematical theory, can be based on set theory.

4.1. In the preceding sections we have emphasized the experience that provable mathematical statements can be formalized by $\{\in\}$-sentences which are derivable from ZFC. Taking this for granted, suppose it were possible in mathematics to prove both a statement and its negation. Let φ be a formalization of this statement. Then both ZFC $\vdash \varphi$ and ZFC $\vdash \neg\varphi$

would hold and thus ZFC would be inconsistent. Therefore, a proof that ZFC is consistent could be regarded as strong evidence for the consistency of mathematics. In fact, the question of the consistency of ZFC is one of the key problems of foundational investigations. In an explicit formulation it asks: Is there a derivation in the sequent calculus of a sequent of the form $\varphi_0 \ldots \varphi_{n-1}(\varphi \wedge \neg\varphi)$, where $\varphi_0, \ldots, \varphi_{n-1}$ are ZFC axioms? In this form, the problem is obviously of a purely syntactic character. Therefore one might hope to solve it by elementary arguments concerning the manipulation of symbol strings by sequent rules. (Hilbert also demanded a proof of such an elementary nature to recognize "that the generally accepted methods of mathematics taken as a whole do not lead to a contradiction".) However, by *Gödel's Second Incompleteness Theorem*, such a consistency proof for ZFC is not possible (cf. X.7). A proof is not even possible if one admits all the auxiliary means of the background set theory described by ZFC.[4] Nevertheless, the fact that ZFC has been investigated and used in mathematics for decades and no inconsistency has been discovered, attests to the consistency of ZFC.

In the following considerations we assume ZFC to be consistent.

4.2. We investigate the relationship between background set theory and object set theory by first discussing *Skolem's paradox* (cf. §2). In terms of ZFC the paradox can be formulated as follows: ZFC, being a countable, consistent set of sentences, has a *countable* model $\mathfrak{A} = (A, \underline{\in}^A)$ according to the Löwenheim–Skolem theorem. On the other hand, $\mathfrak{A}$ satisfies an $\{\underline{\in}\}$-sentence φ (derivable from ZFC) which says that there are *uncountably* many sets in A. If for simplicity we again use defined symbols, we can write

$$\varphi = \exists x \neg \exists y(\underline{\text{Function}}\ y \wedge \underline{\text{Injective}}\ y \wedge \underline{\text{Domain of}}\ y \equiv x \\ \wedge \underline{\text{Range of}}\ y \subseteq \underline{\omega}).$$

φ symbolizes the property of the universe that there exists an uncountable set (and hence also that uncountably many sets exist). Since $\mathfrak{A}$ is a model of ZFC, we have $\mathfrak{A} \models \varphi$, i.e., there is an $a \in A$ (for x) such that

$$(*) \qquad \mathfrak{A} \models \neg \exists y(\underline{\text{Function}}\ y \wedge \cdots \wedge \underline{\text{Range of}}\ y \subseteq \underline{\omega})[a].$$

The set $\{b \in A \mid b \underline{\in}^A a\}$ is at most countable because it is a subset of A. Therefore *in the universe* there exists an injective function whose domain is $\{b \in A \mid b \underline{\in}^A a\}$ and whose range is a subset of ω. This does not contradict $(*)$. For $(*)$ merely says that *in* $\mathfrak{A}$ there is no injective function defined on a with values in $\underline{\omega}^A$, or more exactly, that there is no $b \in A$ such that $\underline{\text{Function}}^A\ b$, $\underline{\text{Injective}}^A\ b$, and $\underline{\text{Domain}}^A\ b \subseteq^A \underline{\omega}^A$; in other words, a is uncountable in $\mathfrak{A}$.

We see from this example that it is necessary to distinguish carefully between the set-theoretical concepts (which refer to the universe) and their meaning in a model.

[4] Since on the basis of background set theory we proved the correctness of the sequent calculus (cf. IV.6.2, 7.5), the preceding remark says in particular that on this basis one cannot show that ZFC is satisfiable, i.e., one cannot prove the existence of a model $(A, \underline{\in}^A)$ of ZFC in the universe.

Let us consider another example. The set of sentences

$$\Psi := \mathrm{ZFC} \cup \{c_r \underset{\sim}{\in} \underset{\sim}{\omega} \mid r \in \mathbb{R}\} \cup \{\neg c_r \equiv c_s \mid r, s \in \mathbb{R}, r \neq s\}$$

is satisfiable, as one can easily show using the compactness theorem. Let $\mathfrak{B} = (B, \underset{\sim}{\in}^B)$ be a model of Ψ (more exactly, the $\{\underset{\sim}{\in}\}$-reduct of a model of Ψ). Then $\{b \in B \mid b \underset{\sim}{\in}^B \underset{\sim}{\omega}^B\}$ is an uncountable set. On the other hand, $\underset{\sim}{\omega}^B$ (being the set of natural numbers in $\mathfrak{B}$) is $\underline{\text{Countable}}^B$ (that is, $\underline{\text{Countable}}^B\, \underset{\sim}{\omega}^B$).

As before let $\mathfrak{A} = (A, \underset{\sim}{\in}^A)$ be a countable model of ZFC. Then $\{a \in A \mid a \underset{\sim}{\in}^A \underset{\sim}{\omega}^A\}$ is countable because it is a subset of A, and we obtain:

(1) There is no bijection of $\{b \in B \mid b \underset{\sim}{\in}^B \underset{\sim}{\omega}^B\}$ onto $\{a \in A \mid a \underset{\sim}{\in}^A \underset{\sim}{\omega}^A\}$,

since one set is uncountable whereas the other one is countable. At first glance (1) seems to contradict Dedekind's theorem, according to which every two Peano structures are isomorphic. To analyze the situation, we take a formalization ψ of this theorem as an $\{\underset{\sim}{\in}\}$-sentence

$$\psi := \forall x\, \forall y((\underline{\text{Peano structure}}\; x \wedge \underline{\text{Peano structure}}\; y) \\ \rightarrow x\; \underline{\text{isomorphic to}}\; y).$$

Then we have

(2) $\mathrm{ZFC} \vdash \psi$.

However, (1) and (2) do not contradict each other. (2) merely says that in each individual model $\mathfrak{C}$ of ZFC every two Peano structures are isomorphic (in the sense of $\mathfrak{C}$), whereas (1) speaks of Peano structures in *different* models.

4.3. We provide a set-theoretic development of first-order logic, i.e., we show that its concepts can be based on the concept of set, as we have done already for functions and Peano structures. To be specific, we restrict ourselves to the symbol set $S = \{P^1, P^2, \ldots\}$, where P^n is n-ary. Our first goal is to give a set-theoretic substitute for S-formulas.

As a substitute for the variables $v_0, v_1, \ldots$ we use the elements $\underset{\sim}{0}, \underset{\sim}{1}, \ldots$ of ω. The rôles of the symbols $\neg, \vee, \exists, \equiv$ are assumed by the ordered pairs $\underset{\sim}{\neg} := (\underset{\sim}{0}, \underset{\sim}{0})$, $\underset{\sim}{\vee} := (\underset{\sim}{0}, \underset{\sim}{1})$, $\underset{\sim}{\exists} := (\underset{\sim}{0}, \underset{\sim}{2})$ and $\underset{\sim}{\equiv} := (\underset{\sim}{0}, \underset{\sim}{3})$, respectively. For the P^n (for $n \geq 1$) we take the ordered pairs $\underset{\sim}{P}^x := (\underset{\sim}{1}, x)$, where $x \in \omega - \{\underset{\sim}{0}\}$. (Similarly, one could, for example, let ordered pairs $(\underset{\sim}{2}, x)$ with $x \in \omega$ stand for function symbols. In order to represent symbol sets with uncountably many elements, one could use an appropriate set of larger cardinality instead of ω.)

Now formulas of the form $v_n \equiv v_m$ correspond to triples $(x, \underset{\sim}{\equiv}, y)$ with $x, y \in \omega$. These triples are the elements of the set

$$At^{\equiv} := \omega \times \{\underset{\sim}{\equiv}\} \times \omega.$$

Ordered pairs of the form $(\underset{\sim}{P}^x, z)$, where $x \in \omega$ and z is a function from x into ω, play the rôle of formulas of the form $P^n v_{m_0} \ldots v_{m_{n-1}}$. (For instance, the formula $P^3 v_1 v_4 v_5$ corresponds to the ordered pair $(\underset{\sim}{P}^3, z)$ with $z = \{(\underset{\sim}{0}, \underset{\sim}{1}), (\underset{\sim}{1}, \underset{\sim}{4}), (\underset{\sim}{2}, \underset{\sim}{5})\}$.) Thus we are led to the set

$$At^P := \{(\underset{\sim}{P}^x, z) \mid x \in \omega \text{ and } z\colon x \to \omega\}.$$

Likewise, one can define the set of all S-formulas set-theoretically to be the smallest set A which satisfies the conditions:

(1) $At^{\equiv} \cup At^P \subset A$;
(2) if $y \in A$ then $(\underset{\sim}{\neg}, y) \in A$;
(3) if $y, z \in A$ then $(y, \underset{\sim}{\vee}, z) \in A$;
(4) if $x \in \omega$ and $y \in A$ then $(\underset{\sim}{\exists}, x, y) \in A$.

One can now give a natural set-theoretic description of the notions of sequent and derivation, developing in this way the whole syntax set-theoretically. Semantic concepts such as the notions of structure or of consequence can also be introduced set-theoretically. By doing this one can obtain a set-theoretic formulation of the completeness theorem. All considerations can be carried out in $L^{\{\in\}}$ on the basis of ZFC. In particular, the completeness theorem can be formalized as an $\{\in\}$-sentence and can be derived from ZFC.

What benefits do we obtain from such a set-theoretical treatment? We mention three points.

(1) The mathematical development of first-order logic (as given in the first six chapters) can be founded upon the axiomatic basis of ZFC.
(2) The set-theoretic treatment enables us to deal with uncountable symbol sets in a precise manner. Appropriate variations of this approach make it possible to define other languages, e.g., languages with infinitely long formulas of the form $\varphi_0 \vee \varphi_1 \vee \varphi_2 \vee \ldots$ (Chapter IX).
(3) In our discussion concerning the formal notion of proof and the scope of first-order logic, we did not appeal to the completeness theorem. The reason for this was to avoid becoming trapped in a vicious circle since the completeness theorem itself requires a proof. In a set-theoretical framework one can investigate more closely the assumptions which are needed for a proof of the completeness theorem. Doing this one finds that a considerably weaker axiom system than ZFC is sufficient for the proof (cf. [1]).

4.4 Exercise. A reader who has been confused by the discussion in this chapter says, "Now I'm completely mixed up. How can ZFC be used as a basis for first-order logic, while first-order logic was actually needed in order to build up ZFC?" Help such a reader out of his dilemma. (*Hint*: Again be careful in distinguishing between the object and the background level.)

CHAPTER VIII

Appendix

In this chapter we note some results which will be needed in Part II and which are also of independent interest.

§1. Extensions by Definitions

We have chosen the symbol set $\{<\}$ for the theory of partially defined orderings and the symbol set $\{\circ, e\}$ for group theory. The field of a partially defined ordering and the inverse operation in a group are examples of relations or functions for which we have no symbol in $\{<\}$ and $\{\circ, e\}$, respectively, although such symbols would facilitate the formulation of propositions. In this section we show that the use of additional symbols does not increase the expressive power of a language provided they are "definable". The field of a partially defined ordering and the inverse operation in a group will turn out to be definable in this sense. For motivation we first discuss these examples in more detail.

When one adds to the symbol set $\{<\}$ a unary relation symbol P for the field of a partially defined ordering one has the following definition:

$$\forall x(Px \leftrightarrow \exists y(x < y \vee y < x)).$$

In $L^{\{<, P\}}$ one can write more transparent formalizations, e.g., the third axiom for the theory of partially defined orderings (cf. III.6.4) may now be written as follows:

$$\forall x \, \forall y((Px \wedge Py) \rightarrow (x < y \vee x \equiv y \vee y < x)).$$

Similarly, the proposition which says that the field of the ordering relation contains at least three elements has a formulation in $L^{\{<, P\}}$ which is both short and easy to read:

$$\exists x \, \exists y \, \exists z (Px \wedge Py \wedge Pz \wedge \neg x \equiv y \wedge \neg x \equiv z \wedge \neg z \equiv y).$$

Thus the introduction of P is convenient for formalizations. On the other hand, it cannot be expected that more propositions about partially defined orderings can be formulated in $L^{\{<, P\}}$ rather than in $L^{\{<\}}$, since every $\{<, P\}$-formula can be transformed into an equivalent $\{<\}$-formula by replacing all subformulas of the form Px by $\exists y (x < y \vee y < x)$.

Similar considerations apply to definable functions and constants: When we add to $S_{\text{gr}} = \{\circ, e\}$ a unary function symbol $^{-1}$ for the group inverse, we have the following definition:

$$\forall x \, \forall y (x^{-1} \equiv y \leftrightarrow x \circ y \equiv e).$$

Note that the right-hand subformula does indeed define a function in every group (regarded as an S_{gr}-structure), since

$$\Phi_{\text{gr}} \models \forall x \, \exists^{=1} y \, x \circ y \equiv e.$$

In a similar way we can introduce the constant $\underline{\varnothing}$ for the empty set in the set theory ZFC by means of the definition

$$\forall x (\underline{\varnothing} \equiv x \leftrightarrow \forall y \, \neg y \,\underline{\in}\, x);$$

for one can show that

$$\text{ZFC} \models \exists^{=1} x \, \forall y \, \neg y \,\underline{\in}\, x.$$

In the sequel $\varphi(v_0, \ldots, v_{n-1})$ stands for a formula φ where the variables occurring free are among $v_0, \ldots, v_{n-1}$.

1.1 Definition. Let Φ be a set of S-sentences.

(a) Let $P \notin S$ be an n-ary relation symbol and $\varphi(v_0, \ldots, v_{n-1})$ an S-formula. Then we say that

$$\forall v_0 \ldots \forall v_{n-1} (P v_0 \ldots v_{n-1} \leftrightarrow \varphi(v_0, \ldots, v_{n-1}))$$

is an *S-definition of P in Φ*.

(b) Let $f \notin S$ be an n-ary function symbol and $\varphi(v_0, \ldots, v_n)$ an S-formula. We say that

$$\forall v_0 \ldots \forall v_{n-1} \, \forall v_n (f v_0 \ldots v_{n-1} \equiv v_n \leftrightarrow \varphi(v_0, \ldots, v_{n-1}, v_n))$$

is an *S-definition of f in Φ* provided that

$$\Phi \models \forall v_0 \ldots \forall v_{n-1} \, \exists^{=1} v_n \, \varphi(v_0, \ldots, v_n).$$

(c) Let $c \notin S$ be a constant and $\varphi(v_0)$ an S-formula. We say that

$$\forall v_0 (c \equiv v_0 \leftrightarrow \varphi(v_0))$$

is an *S-definition of c in Φ* provided that $\Phi \models \exists^{=1} v_0 \, \varphi(v_0)$.

We thus say that $\forall v_0(Pv_0 \leftrightarrow \exists v_1(v_0 < v_1 \vee v_1 < v_0)$ is a $\{<\}$-definition of P in Φ_{pord}, $\forall v_0 \forall v_1(v_0^{-1} \equiv v_1 \leftrightarrow v_0 \circ v_1 \equiv e)$ is an S_{gr}-definition of $^{-1}$ in Φ_{gr}, and $\forall v_0(\underline{\emptyset} \equiv v_0 \leftrightarrow \forall v_1 \neg v_1 \underline{\in} v_0)$ is an $\{\underline{\in}\}$-definition of $\underline{\emptyset}$ in ZFC.

Now assume that a symbol set S and a set Φ of S-sentences are given. Suppose that $S^\Delta \supset S$, and that for every symbol in $S^\Delta - S$ exactly one S-definition in Φ has been fixed, say,

for n-ary $P \in S^\Delta - S$ the S-definition

(i) $\forall v_0 \ldots \forall v_{n-1}(Pv_0 \ldots v_{n-1} \leftrightarrow \varphi_P(v_0, \ldots, v_{n-1}))$,

for n-ary $f \in S^\Delta - S$ the S-definition

(ii) $\forall v_0 \ldots \forall v_{n-1} \forall v_n(fv_0 \ldots v_{n-1} \equiv v_n \leftrightarrow \varphi_f(v_0, \ldots, v_{n-1}, v_n))$,

and for $c \in S^\Delta - S$ the S-definition

(iii) $\forall v_0(c \equiv v_0 \leftrightarrow \varphi_c(v_0))$.

Let Δ be the set of definitions in (i), (ii), and (iii). As in the case of the group inverse, the interpretation of the symbols in $S^\Delta - S$ is determined in every model of Φ:

1.2 Lemma. *If the S-structure $\mathfrak{A}$ is a model of Φ, then there is exactly one S^Δ-structure $\mathfrak{A}^\Delta$ such that*

$$(*) \qquad \mathfrak{A}^\Delta \upharpoonright S = \mathfrak{A} \quad \textit{and} \quad \mathfrak{A}^\Delta \models \Delta.$$

Proof. Suppose $\mathfrak{A}^\Delta \upharpoonright S = \mathfrak{A}$ and $\mathfrak{A}^\Delta \models \Delta$.

Then, by (i),

$$(1) \qquad P^{\mathfrak{A}^\Delta} a_0 \ldots a_{n-1} \quad \text{iff} \quad \mathfrak{A} \models \varphi_P[a_0, \ldots, a_{n-1}]$$

for n-ary $P \in S^\Delta - S$ and $a_0, \ldots, a_{n-1} \in A$;

by (ii),

$$(2) \qquad f^{\mathfrak{A}^\Delta}(a_0, \ldots, a_{n-1}) = a_n \quad \text{iff} \quad \mathfrak{A} \models \varphi_f[a_0, \ldots, a_{n-1}, a_n]$$

for n-ary $f \in S^\Delta - S$ and $a_0, \ldots, a_{n-1}, a_n \in \mathfrak{A}$;

and by (iii),

$$(3) \qquad c^{\mathfrak{A}^\Delta} = a \quad \text{iff} \quad \mathfrak{A} \models \varphi_c[a]$$

for $c \in S^\Delta - S$ and $a \in A$.

(1), (2), and (3) show that there is at most one S^Δ-structure $\mathfrak{A}^\Delta$ satisfying $(*)$. On the other hand, (1), (2), and (3) determine an S^Δ-expansion $\mathfrak{A}^\Delta$ of $\mathfrak{A}$ with the property $\mathfrak{A}^\Delta \models \Delta$. (Indeed, (2) and (3) determine an n-ary function and an element of A since by 1.1(b), (c) we have

$$\mathfrak{A} \models \forall v_0 \ldots \forall v_{n-1} \exists^{=1} v_n \varphi_f(v_0, \ldots, v_{n-1}, v_n)$$

and $\mathfrak{A} \models \exists^{=1} v_0 \varphi_c(v_0)$, respectively.) ☐

It is intuitively obvious that the introduction of defined symbols does not increase the expressive power: For every formula ψ containing defined symbols, one obtains an equivalent formula ψ^∇ without such symbols by replacing the new symbols by their defining formulas. This is the content of

1.3 Theorem on Definitions. *Let $S^\Delta \supset S$ and $\Phi \subset L_0^S$. Let Δ be a set of S-definitions in Φ, one for each symbol in $S^\Delta - S$. Then for each $\psi \in L_n^{S^\Delta}$ there is a formula $\psi^\nabla \in L_n^S$ such that*

(a) *Given an S-structure $\mathfrak{A}$ with $\mathfrak{A} \models \Phi$ and $a_0, \ldots, a_{n-1} \in A$,*

$$\mathfrak{A}^\Delta \models \psi[a_0, \ldots, a_{n-1}] \quad \textit{iff} \quad \mathfrak{A} \models \psi^\nabla[a_0, \ldots, a_{n-1}].$$

($\mathfrak{A}^\Delta$ *denotes the S^Δ-expansion of $\mathfrak{A}$, which, by* 1.2, *is uniquely determined by the condition $\mathfrak{A}^\Delta \models \Delta$.*)

(b) $\Phi \cup \Delta \models \psi \leftrightarrow \psi^\nabla$.

Proof. As already mentioned, we obtain ψ^∇ from ψ by replacing all symbols from $S^\Delta - S$ in ψ by their defining formulas. In this process we must take into account the possibility that a term may contain nested function symbols. Thus, for instance, the formula

$$\psi = \exists x f g x \equiv y$$

(where $f, g \in S^\Delta - S$) will be converted into one of the form

$$\exists x \, \exists u \, \exists v \left(\varphi_g \frac{xu}{v_0 v_1} \wedge \varphi_f \frac{uv}{v_0 v_1} \wedge v \equiv y \right).$$

We define $^\nabla: L^{S^\Delta} \to L^S$ by induction on S^Δ-formulas. The definition for atomic ψ uses induction on $m(\psi)$, where $m(\psi)$ is the number of occurrences of symbols from S^Δ in ψ, including repetitions.

If $m(\psi) = 0$ then $\psi^\nabla := \psi$.

If $m(\psi) > 0$ let $x_0, x_1, x_2, \ldots$ be the enumeration of the variables not occurring in ψ in the order of the enumeration $v_0, v_1, \ldots$.

Case 1. $\psi = P t_0 \ldots t_{n-1}$:

$$\psi^\nabla := \exists x_0 \ldots \exists x_{n-1}((x_0 \equiv t_0)^\nabla \wedge \cdots \wedge (x_{n-1} \equiv t_{n-1})^\nabla \wedge P x_0 \ldots x_{n-1}), \quad \text{if } P \in S,$$

$$\psi^\nabla := \exists x_0 \ldots \exists x_{n-1}\left((x_0 \equiv t_0)^\nabla \wedge \cdots \wedge (x_{n-1} \equiv t_{n-1})^\nabla \wedge \varphi_P \frac{x_0 \ldots x_{n-1}}{v_0 \ldots v_{n-1}}\right), \quad \text{if } P \in S^\Delta - S.$$

(Since P does not occur in $x_i \equiv t_i$, $m(x_i \equiv t_i) < m(\psi)$!)

Case 2. ψ is of the form $t' \equiv t''$. Since $m(\psi) > 0$, ψ is not of the form $x \equiv y$.

(1) $\psi = ft_0 \ldots t_{n-1} \equiv t$:

$$\psi^\nabla := \exists x_0 \ldots \exists x_{n-1} \exists x_n((x_0 \equiv t_0)^\nabla \wedge \cdots \wedge (x_{n-1} \equiv t_{n-1})^\nabla \wedge (x_n \equiv t)^\nabla \wedge fx_0 \ldots x_{n-1} \equiv x_n), \quad \text{if } f \in S,$$

$$\psi^\nabla := \exists x_0 \ldots \exists x_{n-1} \exists x_n\left((x_0 \equiv t_0)^\nabla \wedge \cdots \wedge (x_{n-1} \equiv t_{n-1})^\nabla \wedge (x_n \equiv t)^\nabla \wedge \varphi_f \frac{x_0 \ldots x_n}{v_0 \ldots v_n}\right), \quad \text{if } f \in S^\Delta - S.$$

(2) $\psi = c \equiv t$:

$$\psi^\nabla := \exists x_0((x_0 \equiv t)^\nabla \wedge c \equiv x_0), \quad \text{if } c \in S,$$

$$\psi^\nabla := \exists x_0\left((x_0 \equiv t)^\nabla \wedge \varphi_c \frac{x_0}{v_0}\right), \quad \text{if } c \in S^\Delta - S.$$

(3) $\psi = x \equiv ft_0 \ldots t_{n-1}$ or $\psi = x \equiv c$: analogous to (1) or (2).

This completes the definition for the atomic case.
For the induction step we set

$$(\neg\psi)^\nabla := \neg\psi^\nabla, \qquad (\psi_0 \vee \psi_1)^\nabla := (\psi_0^\nabla \vee \psi_1^\nabla), \quad \text{and} \quad (\exists x\psi)^\nabla := \exists x\psi^\nabla.$$

Now, upon using this definition, it is not difficult to prove by induction that statement (a) of the theorem holds. Furthermore, it is clear that $\psi^\nabla \in L_n^S$ provided $\psi \in L_n^S$. It is also clear that for $\mathfrak{A} \models \Phi$ and $a_0, \ldots, a_{n-1} \in A$

$$(+) \qquad \mathfrak{A}^\Delta \models (\psi \leftrightarrow \psi^\nabla)[a_0, \ldots, a_{n-1}].$$

For (b): Let $\mathfrak{B}$ be an S^Δ-structure such that $\mathfrak{B} \models \Phi \cup \Delta$, and let $b_0, \ldots, b_{n-1} \in B$. We must show that $\mathfrak{B} \models (\psi \leftrightarrow \psi^\nabla)[b_0, \ldots, b_{n-1}]$. For the structure $\mathfrak{A} := \mathfrak{B} \upharpoonright S$, we have $\mathfrak{A}^\Delta = \mathfrak{B}$ by 1.2. Thus (b) follows from (+). □

Given elementarily equivalent S-structures $\mathfrak{A}$ and $\mathfrak{B}$, we have for $\psi \in L_0^{S^\Delta}$:

$$\begin{aligned} \mathfrak{A}^\Delta \models \psi \quad &\text{iff } \mathfrak{A} \models \psi^\nabla \\ &\text{iff } \mathfrak{B} \models \psi^\nabla \\ &\text{iff } \mathfrak{B}^\Delta \models \psi \end{aligned}$$

Hence we have

1.4 Corollary. *If* $\mathfrak{A} \equiv \mathfrak{B}$ *then* $\mathfrak{A}^\Delta \equiv \mathfrak{B}^\Delta$. □

When treating set theory in Chapter VII, we have frequently used defined symbols to obtain more appealing formalizations. Now we have justified this procedure since, by 1.3(b), a formula ψ in the extended language can be regarded as an abbreviation for the $\{\in\}$-formula ψ^∇.

We point out a further application: An ordering is sometimes considered as a structure $(A, <)$, sometimes as a structure $(A, \leq)$. The theorem on definitions implies that the choice of the basic symbols is immaterial, provided the symbols are definable from each other. In the case of orderings we obtain that $L^{\{<\}}$ and $L^{\{\leq\}}$ have the same expressive power. For a precise formulation see exercise 1.11.

A symbol set is called *relational* if it contains only relation symbols. Sometimes it is convenient to be able to restrict oneself to relational symbol sets. Using the theorem on definitions, we show how function symbols and constants can be replaced by relation symbols in order to obtain a relational symbol set. The idea is to consider the graph of a function rather than the function itself.

Let S be a symbol set. For every n-ary $f \in S$ let F be a new $(n + 1)$-ary relation symbol, and for $c \in S$ let C be a new unary relation symbol. Let S^r consist of the relation symbols from S together with the new relation symbols. Thus S^r is relational. We associate with every S-structure $\mathfrak{A}$ an S^r-structure $\mathfrak{A}^r$ by replacing the functions and constants by their graphs:

(1) $A^r := A$;
(2) for $P \in S$, $P^{\mathfrak{A}^r} := P^{\mathfrak{A}}$;
(3) for n-ary $f \in S$ let $F^{\mathfrak{A}^r}$ be the graph of $f^{\mathfrak{A}}$, that is,

$$F^{\mathfrak{A}^r} a_0 \dots a_{n-1} a_n \quad \text{iff} \quad f^{\mathfrak{A}}(a_0, \dots, a_{n-1}) = a_n;$$

(4) for $c \in S$ let $C^{\mathfrak{A}^r}$ be the graph of $c^{\mathfrak{A}}$, that is,

$$C^{\mathfrak{A}^r} a \quad \text{iff } c^{\mathfrak{A}} = a.$$

Correspondingly, one can transform every S-formula into an S^r-formula, replacing atomic subformulas such as $fxy \equiv z$ by $Fxyz$ and $c \equiv x$ by Cx; nested function symbols are treated as in the proof of 1.3. For instance, if $\psi = fcgx \equiv c$ we set $\psi^r = \exists y\, \exists z(Cy \wedge Gxz \wedge Fyzy)$. Then for arbitrary ψ we have that ψ holds in $\mathfrak{A}$ if and only if ψ^r holds in $\mathfrak{A}^r$. A precise statement of this relationship is contained in the following theorem; the proof follows the outline above and is included for the reader who is interested in seeing the details.

1.5 Theorem. *For every $\psi \in L_n^S$ there is $\psi^r \in L_n^{S^r}$ such that for all S-structures $\mathfrak{A}$ and $a_0, \dots, a_{n-1} \in A$:*

$$(+) \qquad \mathfrak{A} \models \psi[a_0, \dots, a_{n-1}] \quad \textit{iff} \quad \mathfrak{A}^r \models \psi^r[a_0, \dots, a_{n-1}].$$

Proof. Instead of proving the theorem directly, we refer to 1.3, now letting $S \cup S^r$ play the rôle of S^Δ and S^r the rôle of S. Further let

$$\Phi := \{\forall v_0 \dots \forall v_{n-1} \exists^{=1} v_n\, Fv_0 \dots v_{n-1} v_n \mid f \in S \text{ and } f \text{ is } n\text{-ary}\} \cup \{\exists^{=1} v_0\, Cv_0 \mid c \in S\},$$

$$\Delta := \{\forall v_0 \dots \forall v_{n-1} \forall v_n(fv_0 \dots v_{n-1} \equiv v_n \leftrightarrow Fv_0 \dots v_{n-1} v_n) \mid f \in S \text{ and } f \text{ is } n\text{-ary}\} \cup \{\forall v_0(c \equiv v_0 \leftrightarrow Cv_0) \mid c \in S\}.$$

For every function symbol f and every constant c in S, Δ contains an S^r-definition in Φ which says that f and c have graphs F and C, respectively. By 1.3, for every formula $\psi \in L_n^{S \cup S^r}$ we have $\psi^\nabla \in L_n^{S^r}$, and for every S^r-structure $\mathfrak{B}$ which satisfies Φ and for all $b_0, \ldots, b_{n-1} \in B$,

$$(*) \qquad \mathfrak{B}^\Delta \models \psi[b_0, \ldots, b_{n-1}] \quad \text{iff } \mathfrak{B} \models \psi^\nabla[b_0, \ldots, b_{n-1}].$$

For $\psi \in L^S$ set $\psi^r := \psi^\nabla$.

Now, in order to prove (+) let $\mathfrak{A}$ be an S-structure. We apply (*) for the case $\mathfrak{B} := \mathfrak{A}^r$ ($\mathfrak{A}^r$ satisfies Φ!). Since $\mathfrak{A}^{r\Delta} \models \Delta$, the transition from $\mathfrak{A}^r$ to $\mathfrak{A}^{r\Delta}$ means that one adds exactly the functions and distinguished elements which were eliminated in passing from $\mathfrak{A}$ to $\mathfrak{A}^r$. Hence $\mathfrak{A}^{r\Delta} \upharpoonright S = \mathfrak{A}$. Thus for $\psi \in L_n^S$ and $a_0, \ldots, a_{n-1} \in A$,

$$\begin{aligned} \mathfrak{A} \models \psi[a_0, \ldots, a_{n-1}] \quad & \text{iff } \mathfrak{A}^{r\Delta} \upharpoonright S \models \psi[a_0, \ldots, a_{n-1}] \\ & \text{iff } \mathfrak{A}^{r\Delta} \models \psi[a_0, \ldots, a_{n-1}] \quad \text{(coincidence lemma)} \\ & \text{iff } \mathfrak{A}^r \models \psi^\nabla[a_0, \ldots, a_{n-1}] \quad \text{(by (*))}, \end{aligned}$$

and since $\psi^\nabla = \psi^r$, the theorem is proved. □

Given S-structures $\mathfrak{A}$ and $\mathfrak{B}$ such that $\mathfrak{A}^r \equiv \mathfrak{B}^r$, we have by 1.4 that $\mathfrak{A}^{r\Delta} \equiv \mathfrak{B}^{r\Delta}$. As $\mathfrak{A}^{r\Delta} \upharpoonright S = \mathfrak{A}$ and $\mathfrak{B}^{r\Delta} \upharpoonright S = \mathfrak{B}$ it follows that $\mathfrak{A} \equiv \mathfrak{B}$. Hence we have obtained

1.6 Corollary. *For any $\mathfrak{A}$ and $\mathfrak{B}$, $\mathfrak{A}^r \equiv \mathfrak{B}^r$ implies $\mathfrak{A} \equiv \mathfrak{B}$.* □

The converse of 1.6 can be shown similarly. Define

$$\begin{aligned} \Delta := \{ & \forall v_0 \ldots \forall v_{n-1} \forall v_n (F v_0 \ldots v_{n-1} v_n \leftrightarrow f v_0 \ldots v_{n-1} \equiv v_n \mid f \in S \text{ is } n\text{-ary}\} \\ & \cup \{\forall v_0 (C v_0 \leftrightarrow c \equiv v_0) \mid c \in S\}. \end{aligned}$$

Then for every new relation symbol in $S^r - S$, Δ contains an S-definition in the empty set of sentences. Clearly for every S-structure $\mathfrak{A}$, $\mathfrak{A}^\Delta \upharpoonright S^r = \mathfrak{A}^r$. Now, if $\mathfrak{A} \equiv \mathfrak{B}$, then 1.4 yields $\mathfrak{A}^\Delta \equiv \mathfrak{B}^\Delta$, and hence by the coincidence lemma we have $\mathfrak{A}^r \equiv \mathfrak{B}^r$. Together with 1.6 we obtain

1.7 Theorem. *For any $\mathfrak{A}$ and $\mathfrak{B}$, $\mathfrak{A} \equiv \mathfrak{B}$ if and only if $\mathfrak{A}^r \equiv \mathfrak{B}^r$.* □

We shall use 1.7 in some discussions on elementary equivalence when it is convenient to restrict to relational symbol sets.

1.8 Exercise. In the notation of 1.3, show that $\Phi \cup \Delta \models \psi$ iff $\Phi \models \psi^\nabla$.

1.9 Exercise. Let us call a formula ψ *term reduced* if its atomic subformulas are of the form $Px_0 \ldots x_{n-1}$, $x \equiv y$, $fx_0 \ldots x_{n-1} \equiv x_n$, or $c \equiv x$. Show that every formula is logically equivalent to a term reduced formula with the same free variables.

1.10 Exercise. Prove theorem 1.5 in the following way: Using 1.9 note that one just needs to consider term reduced formulas. Then give an inductive definition of ψ^r for term reduced formulas.

1.11 Exercise. Show that for any $\varphi \in L_0^{\{<\}}$ there is a $\psi \in L_0^{\{\leq\}}$ and for any $\psi \in L_0^{\{\leq\}}$ there is a $\varphi \in L_0^{\{<\}}$ such that

(a) an ordering $(A, <^A)$ satisfies φ if and only if the corresponding ordering $(A, \leq^A)$ satisfies ψ,
(b) an ordering $(A, \leq^A)$ satisfies ψ if and only if the corresponding ordering $(A, <^A)$ satisfies φ.

§2. Relativization and Substructures

If one regards a vector space as a one-sorted structure, then the domain consists of scalars and vectors (cf. III.7.2(2)). When formulating the vector space axioms in the corresponding language, one must *relativize* the field axioms to the set of scalars and the group axioms (for the vectors) to the set of vectors. For the field axiom $\forall x\, x \cdot 1 \equiv x$ this can be done by using the relation symbol $\underline{F}$ for the set of scalars and reformulating the axiom as $\forall x(\underline{F}x \rightarrow x \cdot 1 \equiv x)$. Similarly, the S_{ar}-formula

$$\varphi := \forall x(x \equiv 0 \vee x \equiv 1),$$

when relativized to $\underline{F}$, becomes

$$\varphi^{\underline{F}} := \forall x(\underline{F}x \rightarrow (x \equiv 0 \vee x \equiv 1)).$$

In a vector space, $\varphi^{\underline{F}}$ just says that the field of scalars satisfies φ. This section deals with the relation between a formula and its relativization. First it is useful to introduce the notion of substructure.

2.1 Definition. Let $\mathfrak{A}$ and $\mathfrak{B}$ be S-structures. Then $\mathfrak{A}$ is called a *substructure* of $\mathfrak{B}$ (written: $\mathfrak{A} \subset \mathfrak{B}$) if

(a) $A \subset B$;
(b) (1) for n-ary $P \in S$, $P^{\mathfrak{A}} = P^{\mathfrak{B}} \cap A^n$ (that is, for all $a_0, \ldots, a_{n-1} \in A$, $P^{\mathfrak{A}}a_0 \ldots a_{n-1}$ iff $P^{\mathfrak{B}}a_0 \ldots a_{n-1}$);
(2) for n-ary $f \in S$, $f^{\mathfrak{A}}$ is the restriction of $f^{\mathfrak{B}}$ to A^n;
(3) for $c \in S$, $c^{\mathfrak{A}} = c^{\mathfrak{B}}$.

For example, $\mathfrak{N} = (\mathbb{N}, +^{\mathbb{N}}, \cdot^{\mathbb{N}}, 0, 1)$ is a substructure of the field $\mathfrak{R} = (\mathbb{R}, +^{\mathbb{R}}, \cdot^{\mathbb{R}}, 0, 1)$ of real numbers.

If $\mathfrak{A} \subset \mathfrak{B}$, then A is S-*closed* (in $\mathfrak{B}$), that is, $A \neq \varnothing$, $c^{\mathfrak{B}} \in A$ for $c \in S$, and $a_0, \ldots, a_{n-1} \in A$ implies $f^{\mathfrak{B}}(a_0, \ldots, a_{n-1}) \in A$ for $f \in S$.

Conversely, every S-closed subset X of B is the domain of exactly one substructure of $\mathfrak{B}$, because in this case the conditions in 2.1(b) determine exactly one structure with domain X. We denote this substructure by $[X]^{\mathfrak{B}}$.

For example, the set $\{r \in \mathbb{R} \mid r \geq 0\}$ is S_{ar}-closed in $\mathfrak{R}$ and hence is the domain of a substructure of $\mathfrak{R}$, but the set $\{r \in \mathbb{R} \mid r \leq 0\}$ is not S_{ar}-closed, since $(-1)\cdot(-1)$ is not ≤ 0.

2.2 Lemma. *Let $\mathfrak{A}$ and $\mathfrak{B}$ be S-structures such that $\mathfrak{A} \subset \mathfrak{B}$, and let $\beta\colon \{v_n \mid n \in \mathbb{N}\} \to A$ be an assignment. Then, for every S-term t and for every atomic S-formula φ,*

$$(\mathfrak{A}, \beta)(t) = (\mathfrak{B}, \beta)(t), \qquad (\mathfrak{A}, \beta) \models \varphi \quad \textit{iff} \ (\mathfrak{B}, \beta) \models \varphi.$$

The proof by induction on terms is straightforward. □

The result 2.2 does not hold for arbitrary φ: the formula $\exists x\, x + 1 \equiv 0$ holds in $\mathfrak{R}$ but not in its substructure $\mathfrak{N}$.

2.3 Definition. Let S be a symbol set and let P be a unary relation symbol. By induction we define, for every $\psi \in L^S$, the formula $\psi^P \in L^{S \cup \{P\}}$, the so-called *P-relativization* of ψ:

$$\begin{aligned} \psi^P &:= \psi, \quad \text{if } \psi \text{ is atomic;} \\ [\neg\psi]^P &:= \neg\psi^P; \\ (\psi_0 \vee \psi_1)^P &:= (\psi_0^P \vee \psi_1^P); \\ [\exists x\psi]^P &:= \exists x(Px \wedge \psi^P). \end{aligned}$$

Thus $[\forall x\psi]^P = [\neg\exists x \neg\psi]^P = \neg\exists x(Px \wedge \neg\psi^P)$, and this formula is logically equivalent to $\forall x(Px \to \psi^P)$. Moreover, it is clear that $\text{free}(\psi) = \text{free}(\psi^P)$.

2.4 Relativization Lemma. *Let $\mathfrak{A}$ be an $S \cup \{P\}$-structure such that $P^A \subset A$ is an S-closed set. Then for $\psi \in L_0^S$,*

$$[P^A]^{\mathfrak{A}} \models \psi \quad \textit{iff} \ \mathfrak{A} \models \psi^P.$$

2.4 describes the relationship sketched at the beginning of this section: the relativization ψ^P says the same in $\mathfrak{A}$ as ψ does in the substructure $[P^A]^{\mathfrak{A}}$.

Proof. We show by induction on $\psi \in L^S$:

(*) For all assignments $\beta\colon \{v_n \mid n \in \mathbb{N}\} \to P^A$,

$$([P^A]^{\mathfrak{A}}, \beta) \models \psi \quad \text{iff} \ (\mathfrak{A}, \beta) \models \psi^P.$$

If ψ is atomic, then $\psi^P = \psi$, and we obtain (*) from 2.2. For $\psi = \neg\psi_0$ or $\psi = (\psi_0 \vee \psi_1)$, (*) follows directly from the induction hypothesis for ψ_0 and ψ_1. In case $\psi = \exists x\psi_0$ we argue as follows:

$$([P^A]^{\mathfrak{A}}, \beta) \models \exists x\psi_0 \quad \text{iff for some } a \in P^A, \left([P^A]^{\mathfrak{A}}, \beta\frac{a}{x}\right) \models \psi_0$$

$$\text{iff for some } a \in P^A, \left(\mathfrak{A}, \beta\frac{a}{x}\right) \models \psi_0^P \quad \text{(by induction hypothesis for } \psi_0\text{)}$$

$$\text{iff for some } a \in A, \left(\mathfrak{A}, \beta\frac{a}{x}\right) \models Px \wedge \psi_0^P$$

$$\text{iff } (\mathfrak{A}, \beta) \models \exists x(Px \wedge \psi_0^P).$$

□

2.5 Exercise. Let U and V be distinct unary relation symbols, $U, V \notin S$. Assume $(\mathfrak{A}, U^A, V^A)$ to be an $S \cup \{U, V\}$-structure such that U^A and V^A are S-closed and $U^A \subset V^A$. Show that for $\varphi \in L_0^S$

$$(\mathfrak{A}, U^A, V^A) \models ([\varphi^V]^U \leftrightarrow \varphi^U).$$

2.6 Exercise. A formula of the form $\exists x_0 \ldots \exists x_{n-1}\varphi$ $(\forall x_0 \ldots \forall x_{n-1}\varphi)$, where $n \geq 0$ and φ does not contain any quantifiers, is called an existential formula (universal formula). Show:

(a) If $\mathfrak{A} \subset \mathfrak{B}$, φ is an existential sentence and $\mathfrak{A} \models \varphi$, then $\mathfrak{B} \models \varphi$.
(b) If $\mathfrak{A} \subset \mathfrak{B}$, φ is a universal sentence and $\mathfrak{B} \models \varphi$, then $\mathfrak{A} \models \varphi$.
(c) In the language $L^{S_{\text{gr}}}$ there is no system of axioms for group theory consisting only of universal sentences.

§3. Normal Forms

In this section we show that one can associate with every formula a logically equivalent formula which has a special syntactic form.

Let S be a fixed symbol set. For an arbitrary set Φ of S-formulas let $\langle\Phi\rangle$ be the smallest subset of L^S containing Φ, and containing with any ψ and χ the formulas $\neg\psi$ and $(\psi \vee \chi)$. Note that $\Phi \subset L_r^S$ implies $\langle\Phi\rangle \subset L_r^S$.

3.1 Lemma. *Let $\Phi \subset L_r^S$. Suppose $\mathfrak{A}$ and $\mathfrak{B}$ are S-structures, and $a_0, \ldots, a_{r-1} \in A$, $b_0, \ldots, b_{r-1} \in B$. If*

$$(*) \qquad \mathfrak{A} \models \varphi[a_0, \ldots, a_{r-1}] \quad \textit{iff } \mathfrak{B} \models \varphi[b_0, \ldots, b_{r-1}]$$

holds for all $\varphi \in \Phi$, then (*) holds for all $\varphi \in \langle\Phi\rangle$.

Proof. The set of φ for which (*) holds includes Φ and with any ψ and χ also contains $\neg\psi$ and $(\psi \vee \chi)$. □

3.2 Lemma. *Let $\Phi = \{\varphi_0, \ldots, \varphi_n\}$ be a finite set of formulas. Then every satisfiable formula in $\langle\Phi\rangle$ is logically equivalent to a formula of the form*

$$(+) \qquad (\psi_{0,0} \wedge \cdots \wedge \psi_{0,n}) \vee \cdots \vee (\psi_{k,0} \wedge \cdots \wedge \psi_{k,n}),$$

where $k < 2^{n+1}$ and for $i \leq k$ and $j \leq n$, $\psi_{i,j} = \varphi_j$ or $\psi_{i,j} = \neg\varphi_j$. In particular, there are only finitely many pairwise logically nonequivalent formulas in $\langle\Phi\rangle$.

Thus we see that every formula in $\langle\Phi\rangle$ is logically equivalent to a disjunction of conjunctions of formulas from $\{\varphi_0, \ldots, \varphi_n, \neg\varphi_0, \ldots, \neg\varphi_n\}$.

PROOF. We choose an r such that $\Phi = \{\varphi_0, \ldots, \varphi_n\} \subset L_r^S$. For a structure $\mathfrak{A}$ and an r-tuple $\mathring{a} := (a_0, \ldots, a_{r-1}) \in A^r$ let

$$(1) \qquad \psi_{(\mathfrak{A}, \mathring{a})} := \psi_0 \wedge \cdots \wedge \psi_n,$$

where

$$\psi_i := \begin{cases} \varphi_i, & \text{if } \mathfrak{A} \models \varphi_i[a_0, \ldots, a_{r-1}], \\ \neg\varphi_i, & \text{if } \mathfrak{A} \models \neg\varphi_i[a_0, \ldots, a_{r-1}]. \end{cases}$$

Then

$$(2) \qquad \mathfrak{A} \models \psi_{(\mathfrak{A}, \mathring{a})}[a_0, \ldots, a_{r-1}],$$

and $\psi_{(\mathfrak{A}, \mathring{a})}$ is a conjunction of the form in (+). Moreover, for any $\mathfrak{B}$ and $b_0, \ldots, b_{r-1} \in \mathfrak{B}$,

$$(3) \qquad \mathfrak{B} \models \psi_{(\mathfrak{A}, \mathring{a})}[b_0, \ldots, b_{r-1}]$$

iff for $i = 0, \ldots, n$:

$$\mathfrak{A} \models \varphi_i[a_0, \ldots, a_{r-1}] \quad \text{iff } \mathfrak{B} \models \varphi_i[b_0, \ldots, b_{r-1}]$$

iff (cf. 3.1) for all $\varphi \in \langle\Phi\rangle$:

$$\mathfrak{A} \models \varphi[a_0, \ldots, a_{r-1}] \quad \text{iff } \mathfrak{B} \models \varphi[b_0, \ldots, b_{r-1}].$$

From (1) it follows that the set

$$\{\psi_{(\mathfrak{A}, \mathring{a})} \mid \mathfrak{A} \text{ is an } S\text{-structure and } \mathring{a} \in A^r\}$$

has at most 2^{n+1} elements.

The proof is complete if we can show that every satisfiable $\varphi \in \langle\Phi\rangle$ is logically equivalent to the disjunction χ of the finitely many formulas from the set

$$\{\psi_{(\mathfrak{A}, \mathring{a})} \mid \mathfrak{A} \text{ is an } S\text{-structure, } \mathring{a} \in A^r, \mathfrak{A} \models \varphi[a_0, \ldots, a_{r-1}]\}.$$

In a suggestive notation we write

$$\chi = \bigvee\{\psi_{(\mathfrak{A}, \mathring{a})} \mid \mathfrak{A} \text{ is an } S\text{-structure, } \mathring{a} \in A^r, \mathfrak{A} \models \varphi[a_0, \ldots, a_{r-1}]\}.$$

To verify the equivalence between φ and χ, assume first that $\mathfrak{B} \models \varphi[b_0, \ldots, b_{r-1}]$. Then $\psi_{(\mathfrak{B}, \vec{b})}$ is a member of the disjunction χ. Since $\mathfrak{B} \models \psi_{(\mathfrak{B}, \vec{b})}[b_0, \ldots, b_{r-1}]$ (cf. (2)) it follows that $\mathfrak{B} \models \chi[b_0, \ldots, b_{r-1}]$. Conversely, if $\mathfrak{B} \models \chi[b_0, \ldots, b_{r-1}]$, then by definition of χ there is a structure $\mathfrak{A}$ and there are $a_0, \ldots, a_{r-1} \in A$ such that

$$\mathfrak{A} \models \varphi[a_0, \ldots, a_{r-1}] \quad \text{and} \quad \mathfrak{B} \models \psi_{(\mathfrak{A}, \vec{a})}[b_0, \ldots, b_{r-1}].$$

Then by (3), $b_0, \ldots, b_{r-1}$ satisfy the same formulas of $\langle\Phi\rangle$ in $\mathfrak{B}$ as $a_0, \ldots, a_{r-1}$ do in $\mathfrak{A}$. In particular, $\mathfrak{B} \models \varphi[b_0, \ldots, b_{r-1}]$. □

A formula which is a disjunction of conjunctions of atomic or negated atomic formulas is called a *formula in disjunctive normal form.* A formula which contains no quantifiers is said to be *quantifier-free.* As a corollary to 3.2 we obtain

3.3 Theorem on the Disjunctive Normal Form. *If φ is quantifier-free, then φ is logically equivalent to a formula in disjunctive normal form.*

PROOF. Let φ be a quantifier-free formula. If φ is not satisfiable then φ is logically equivalent to $\neg v_0 \equiv v_0$. If φ is satisfiable and $\psi_0, \ldots, \psi_n$ are the atomic subformulas in φ, then $\varphi \in \langle\{\psi_0, \ldots, \psi_n\}\rangle$. The theorem now follows from 3.2. □

We turn to formulas which also contain quantifiers. A formula ψ is said to be in *prenex normal form* if it has the form $Q_0 x_0 \ldots Q_{m-1} x_{m-1} \psi_0$, where $Q_i = \exists$ or $Q_i = \forall$ for $i < m$ and ψ_0 is quantifier-free. $Q_0 x_0 \ldots Q_{m-1} x_{m-1}$ is called the *prefix* and ψ_0 the *matrix* of ψ.

3.4 Theorem on the Prenex Normal Form. *Every formula φ is logically equivalent to a formula ψ in prenex normal form with* free(φ) = free(ψ).

PROOF. First we note some simple properties of logical equivalence. By "$\varphi \sim \psi$" we mean that φ and ψ are logically equivalent.

(1) If $\varphi \sim \psi$, then $\neg\varphi \sim \neg\psi$.
(2) If $\varphi_0 \sim \psi_0$ and $\varphi_1 \sim \psi_1$, then $(\varphi_0 \vee \varphi_1) \sim (\psi_0 \vee \psi_1)$.
(3) If $\varphi \sim \psi$ and $Q = \exists$ or $Q = \forall$, then $Qx\varphi \sim Qx\psi$.
(4) $\neg\exists x\varphi \sim \forall x \neg\varphi$, $\neg\forall x\varphi \sim \exists x \neg\varphi$.
(5) If $x \notin \text{free}(\psi)$, then $(\exists x\varphi \vee \psi) \sim \exists x(\varphi \vee \psi)$, $(\forall x\varphi \vee \psi) \sim \forall x(\varphi \vee \psi)$, $(\psi \vee \exists x\varphi) \sim \exists x(\psi \vee \varphi)$, and $(\psi \vee \forall x\varphi) \sim \forall x(\psi \vee \varphi)$.

We shall see how one can transform a given formula into prenex normal form by repeated application of (1)–(5). For instance, if $\varphi = \neg\exists xPx \vee \forall xRx$ we can proceed as follows:

$$\begin{aligned} \neg\exists xPx \vee \forall xRx &\sim \forall x\neg Px \vee \forall xRx && \text{(by (2) and (4))} \\ &\sim \forall x\neg Px \vee \forall yRy && \text{(since } \forall xRx \sim \forall yRy \text{ and by (2))} \\ &\sim \forall x(\neg Px \vee \forall yRy) && \text{(by (5))} \\ &\sim \forall x\,\forall y(\neg Px \vee Ry) && \text{(by (3) and (5)).} \end{aligned}$$

In general we argue as follows: For $\varphi \in L^S$ let $\text{qn}(\varphi)$ be the *quantifier number* of φ, i.e., the number of quantifiers occurring in φ. Using induction on n, we prove:

$(*)_n$ For φ with $\text{qn}(\varphi) \leq n$ there is a $\psi \in L^S$ in prenex normal form such that $\varphi \sim \psi$, $\text{free}(\varphi) = \text{free}(\psi)$, and $\text{qn}(\varphi) = \text{qn}(\psi)$.

We leave the arguments for "$\text{free}(\varphi) = \text{free}(\psi)$" to the reader.

$n = 0$: If $\text{qn}(\varphi) = 0$, φ is quantifier-free and we can set $\psi := \varphi$.

$n > 0$: We show $(*)_n$ by induction on φ. Suppose $\text{qn}(\varphi) \leq n$. The quantifier-free case is clear. If $\varphi = \neg\varphi'$ and $\text{qn}(\varphi) > 0$, then $\text{qn}(\varphi') = \text{qn}(\varphi) > 0$, and by induction hypothesis there is a formula of the form $Qx\chi$ which is a prenex normal form for φ' (where $\text{qn}(Qx\chi) = \text{qn}(\varphi)$ and where χ may contain quantifiers). By (1) and (4), $\varphi \sim Q^{-1}x\neg\chi$ (where $\forall^{-1} := \exists$ and $\exists^{-1} := \forall$). Since $\text{qn}(\neg\chi) = \text{qn}(Qx\chi) - 1 = \text{qn}(\varphi) - 1 \leq n - 1$, there exists a formula ψ logically equivalent to $\neg\chi$ which is in prenex normal form such that $\text{qn}(\psi) = \text{qn}(\neg\chi)$. By (3), $Q^{-1}x\psi$ is a formula logically equivalent to φ with the desired properties.

Let $\varphi = (\varphi' \vee \varphi'')$ and let $\text{qn}(\varphi) > 0$, e.g., $\text{qn}(\varphi') > 0$. By induction hypothesis there is a formula of the form $Qx\chi$ which is a prenex normal form for φ'. Let y be a variable which does not occur in $Qx\chi$ or in φ''. It is easy to show that

$$Qx\chi \sim Qy\chi\frac{y}{x}$$

and thus, by (2) and (5), to obtain

$$\begin{aligned}\varphi = (\varphi' \vee \varphi'') &\sim \left(Qy\chi\frac{y}{x} \vee \varphi''\right)\\ &\sim Qy\left(\chi\frac{y}{x} \vee \varphi''\right).\end{aligned}$$

Since $\text{qn}(\chi\frac{y}{x} \vee \varphi'') = \text{qn}(\varphi) - 1 \leq n - 1$, we can find a formula ψ in prenex normal form which is logically equivalent to $(\chi\frac{y}{x} \vee \varphi'')$. $Qy\psi$ has the desired properties.

Let $\varphi = \exists x\varphi'$. Since $\text{qn}(\varphi') \leq n - 1$ there is a formula ψ' in prenex normal form which is logically equivalent to φ'. $\exists x\psi'$ is a formula in prenex normal form which, by (3), is logically equivalent to φ and has the same quantifier number as φ. □

3.5 Exercise (Conjunctive Normal Form). Show that if φ is quantifier-free, then φ is logically equivalent to a formula which is a conjunction of disjunctions of atomic and negated atomic formulas.

3.6 Exercise. Let S be a relational symbol set and let $\varphi \in L_0^S$ be of the form $\exists x_0 \ldots \exists x_n \forall y_0 \ldots \forall y_m \psi$, where ψ is quantifier free. Show that every model of φ contains an $(n + 1)$-element substructure which is also a model of φ. Conclude that the sentence $\forall x \exists y Rxy$ is not equivalent to any $\{R\}$-sentence of the above form.

PART B

CHAPTER IX

Extensions of First-Order Logic

We have seen that the structure $\mathfrak{N}$ of natural numbers cannot be characterized in the first-order language corresponding to $\mathfrak{N}$. The same situation holds for the class of torsion groups. As we showed in Chapter VII, one can, at least in principle, overcome this weakness by a set-theoretical formulation: One introduces a system of axioms for set theory in a first-order language, e.g., ZFC, which is sufficient for mathematics, and then in this system carries out the arguments which are required, say, for a definition and characterization of $\mathfrak{N}$. However, this approach necessitates explicit use of set theory to an extent not usual in ordinary mathematical practice.

The situation may act as a motivation to consider languages with more expressive power which permit us to avoid this detour through set theory. For example, in a second-order language we can directly characterize the natural numbers by means of Peano's axioms. However, already at this stage we wish to remark that in order to set up the semantics of such a language and to prove the correctness of inference rules, one has to make more extensive use of set-theoretic assumptions (for example, of the ZFC axioms) than for first-order logic.

There is a further reason for introducing and investigating more powerful languages. We saw that results such as the compactness theorem are useful in algebraic investigations (cf. VI.4). Therefore it seems worthwhile to seek other more expressive languages in the hope of obtaining tools for more far-reaching applications in mathematics.

In this chapter we introduce the reader to some of the languages which have been considered with these aims in mind.

§1. Second-Order Logic

The difference between second-order and first-order languages lies in the fact that in the former one can quantify over second-order objects (for example, subsets of the domain of a structure) whereas in the latter this is not possible.

1.1 The Second-Order Language L_{II}^S. Let S be a symbol set, that is, a set of relation symbols, function symbols and constants. The alphabet of L_{II}^S contains, in addition to the symbols of L^S, for each $n \geq 1$ countably many n-ary relation variables $V_0^n, V_1^n, V_2^n, \ldots$. To denote relation variables we use letters $X, Y, \ldots$. We define the set L_{II}^S of second-order S-formulas to be the set generated by the rules of the calculus for first-order formulas (cf. II.3.2), extended by the following two rules:

(a) If X is an n-ary relation variable and $t_0, \ldots, t_{n-1}$ are S-terms, then $Xt_0 \ldots t_{n-1}$ is an S-formula.
(b) If φ is an S-formula and X is a relation variable, then $\exists X\varphi$ is an S-formula.

1.2 The Satisfaction Relation for L_{II}^S. A *second-order assignment* γ in a structure $\mathfrak{A}$ is a map which assigns to each variable v_i an element of A and to each relation variable V_i^n an n-ary relation on A. We extend the notion of satisfaction from L^S to L_{II}^S by taking (a) and (b) into account as follows: If $\mathfrak{A}$ is an S-structure, γ a second-order assignment in $\mathfrak{A}$, and $\mathfrak{I} = (\mathfrak{A}, \gamma)$, then we set

(a′) $\mathfrak{I} \models Xt_0 \ldots t_{n-1}$ iff $\gamma(X)$ holds for $\mathfrak{I}(t_0), \ldots, \mathfrak{I}(t_{n-1})$.

(b′) $\mathfrak{I} \models \exists X\varphi$ iff there is a $C \subset A^n$ such that $\mathfrak{I}\frac{C}{X} \models \varphi$ $\left(\text{where } \mathfrak{I}\frac{C}{X} = \left(\mathfrak{A}, \gamma\frac{C}{X}\right) \text{ and } \gamma\frac{C}{X} \text{ is the assignment which maps } X \text{ to } C \text{ but which otherwise agrees with } \gamma\right)$.

We write $\mathscr{L}_{\mathrm{II}}$ to denote *second-order logic*, that is, the logical system given by the languages L_{II}^S together with the satisfaction relation for these languages. Similarly, we denote first-order logic by $\mathscr{L}_{\mathrm{I}}$. For the present we still use the term "logical system" in the naive sense. A precise definition will be given in XII.1.

1.3 Remarks and Examples

(1) One defines the free occurrence of variables and relation variables in second-order formulas in the obvious way and can then prove the analogue of the coincidence lemma. In particular, when φ is an L_{II}^S-sentence, i.e., a formula without free variables or free relation variables, it is meaningful to say that $\mathfrak{A}$ is a model of φ, written $\mathfrak{A} \models \varphi$.

(2) Let $\forall X\varphi$ be an abbreviation for $\neg\exists X\neg\varphi$. Then

$$\mathfrak{I} \models \forall X\varphi \quad \text{iff for all } C \subset A^n\text{: } \mathfrak{I}\frac{C}{X} \models \varphi.$$

(3) If X is a unary relation variable, then the following formalizations of Peano's axioms,

(P1) $\forall x \neg \underline{\sigma}x \equiv 0$;
(P2) $\forall x \forall y(\underline{\sigma}x \equiv \underline{\sigma}y \rightarrow x \equiv y)$;
(P3) $\forall X((X0 \wedge \forall x(Xx \rightarrow X\underline{\sigma}x)) \rightarrow \forall y Xy)$,

which we had in III.7.3, are $L_{\mathrm{II}}^{\{\sigma, 0\}}$-sentences. Hence by passing from first-order to second-order logic we have gained expressive power since no first-order axioms can characterize the structure $(\mathbb{N}, \sigma, 0)$ up to isomorphism.

(4) Let S be arbitrary. Then the L_{II}^S-sentence

$$(+) \qquad \forall x \forall y(x \equiv y \leftrightarrow \forall X(Xx \leftrightarrow Xy))$$

is valid: two things are equal precisely when there is no property which distinguishes them (the *identitas indiscernibilium* of Leibniz). Thus in the development of L_{II}^S we could have done without the equality symbol using (+) to express equality.

(5) When setting up the second-order languages we could have introduced, in addition to relation variables, *function variables* which can also be quantified. This procedure would increase convenience, but not the expressive power of the languages. We illustrate this by means of an example (cf. the elimination of function symbols in VIII.1).

Let g be a unary function variable and let φ be the "second-order formula"

$$\forall g(\forall x \forall y(gx \equiv gy \rightarrow x \equiv y) \rightarrow \forall x \exists y\, x \equiv gy).$$

Then (for the natural extension of the notion of satisfaction) the following holds:

$$\mathfrak{A} \models \varphi \quad \text{iff every injective function from } A \text{ to } A \text{ is surjective}$$
$$\text{iff } A \text{ is finite.}$$

Considering the graph of a unary function instead of the function itself, we can use a binary relation variable and rewrite φ as

$$\forall X((\forall x \exists^{=1} y\, Xxy \wedge \forall x \forall y \forall z((Xxz \wedge Xyz) \rightarrow x \equiv y)) \rightarrow \forall x \exists y\, Xyx).$$

Call this formula φ_{fin}. φ and φ_{fin} have the same models. Therefore,

$$\mathfrak{A} \models \varphi_{\mathrm{fin}} \quad \text{iff } A \text{ is finite.}$$

In later examples we shall often use function variables to obtain formulas which are easier to read.

(6) In $\mathscr{L}_{\mathrm{II}}$ one can introduce operations such as substitution and relativization (cf. VIII.2.3) by definitions analogous to those for $\mathscr{L}_{\mathrm{I}}$. One can also verify basic semantic properties such as the analogue of the isomorphism lemma.

The situation is different when we consider deeper semantic properties such as the completeness theorem, the compactness theorem and the Löwenheim–Skolem theorem: the price we have to pay for being able to quantify over second-order objects is the loss of all these central properties.

1.4 Theorem. *The compactness theorem does not hold for* $\mathscr{L}_{\mathrm{II}}$.

PROOF. The following set of sentences is a counterexample:

$$\{\varphi_{\mathrm{fin}}\} \cup \{\varphi_{\geq n} \mid n \geq 2\}.$$

This set is not satisfiable, but, of course, every finite subset is satisfiable. □

1.5 Theorem. *The Löwenheim–Skolem theorem does not hold for* $\mathscr{L}_{\mathrm{II}}$.

PROOF. We give a sentence $\varphi_{\mathrm{unc}} \in L_{\mathrm{II}}^{\varnothing}$ such that for all structures $\mathfrak{A}$,

$$\mathfrak{A} \models \varphi_{\mathrm{unc}} \quad \text{iff } A \text{ is uncountable.}$$

Then φ_{unc} is satisfiable but it has no model that is at most countable.

To define φ_{unc} we use an $L_{\mathrm{II}}^{\varnothing}$-formula $\varphi_{\mathrm{fin}}(X)$ with just one free unary relation variable X, for which

$$(\mathfrak{A}, \gamma) \models \varphi_{\mathrm{fin}}(X) \quad \text{iff } \gamma(X) \text{ is finite.}$$

It is easy to obtain such a formula by modifying φ_{fin} as given above. Clearly a set A is at most countable if and only if there is an ordering relation on A such that every element has only finitely many predecessors. So let us define, using a binary relation variable Y,

$$\begin{aligned} \varphi_{\leq \mathrm{ctbl}} := \exists Y(&\forall x \neg Yxx \wedge \forall x \forall y \forall z((Yxy \wedge Yyz) \rightarrow Yxz) \\ &\wedge \forall x \forall y(Yxy \vee x \equiv y \vee Yyx) \\ &\wedge \forall x \exists X(\varphi_{\mathrm{fin}}(X) \wedge \forall y(Xy \leftrightarrow Yyx))). \end{aligned}$$

Then we have

$$\mathfrak{A} \models \varphi_{\leq \mathrm{ctbl}} \quad \text{iff } A \text{ is at most countable}$$

and hence we can set $\varphi_{\mathrm{unc}} := \neg \varphi_{\leq \mathrm{ctbl}}$. □

1.6. For first-order logic we obtained the compactness theorem from the existence of an adequate system of derivation rules (cf. VI.2). *For* $\mathscr{L}_{\mathrm{II}}$ *there is no correct and complete system of derivation rules.* Otherwise we could use the same argument as for $\mathscr{L}_{\mathrm{I}}$ to prove the compactness theorem for $\mathscr{L}_{\mathrm{II}}$.

This negative result does not, of course, hinder us from setting up correct rules for second-order logic. For example, one can add to the first-order rules the following correct rules for quantification over relation variables:

$$\frac{\Gamma \quad \varphi}{\Gamma \quad \exists X\varphi}; \qquad \frac{\Gamma \qquad \varphi \quad \psi}{\Gamma \quad \exists X\varphi \quad \psi}, \quad \text{if } X \text{ is not free in } \Gamma \; \psi.$$

In the introduction to this chapter we provided two motivations for investigating more expressive languages, namely: (a) to facilitate the formalization of mathematical statements and arguments, and (b) to supply us with more powerful tools for mathematical investigations. In regard to (a) and (b), what have we accomplished by second-order logic?

To begin with, we note that by supplementing the second-order rules presented above, one can obtain a system largely sufficient for the purposes of mathematics. (However, by 1.6, one never gets a complete system, so that the choice of rules can only be made from a pragmatic point of view, and not with the aim of attaining completeness.) In addition, bearing in mind that mathematics can be formulated more conveniently in a second-order language, one can tend to the opinion that progress in the sense of (a) has indeed been made. However, as far as (b) is concerned $\mathscr{L}_{\mathrm{II}}$ is hardly an appropriate system. The results 1.4 and 1.5 already hint at this. The expressive power of second-order languages is so great that results such as the compactness theorem or the Löwenheim–Skolem theorem, which are of value for mathematical applications, no longer hold. In view of these remarks it is natural to investigate other extensions of first-order logic (cf. §2, §3).

By considering a further aspect we explain how, in a certain sense, second-order logic has overshot the mark: We show that set theory, as based, e.g., on ZFC, is not sufficient to decide basic semantic questions for $\mathscr{L}_{\mathrm{II}}$. This follows because we can write down a sentence $\varphi_{\mathrm{CH}} \in L_{\mathrm{II}}^{\varnothing}$ which is valid if and only if the continuum hypothesis CH holds. Since neither CH nor its negation can be proved in ZFC (cf. VII.3), the validity of φ_{CH} can neither be established nor refuted within the framework of ZFC.

CH says:

(1) For every subset A of $\mathbb{R}$, either A is at most countable, or there is a bijection of $\mathbb{R}$ onto A.

φ_{CH} will be essentially a formalization of (1). First, similar to $\varphi_{\leq \mathrm{ctbl}}$, we can easily give a formula $\varphi_{\leq \mathrm{ctbl}}(X)$ with the property

$$(\mathfrak{A}, \gamma) \models \varphi_{\leq \mathrm{ctbl}}(X) \quad \text{iff} \quad \gamma(X) \text{ is at most countable.}$$

Further, there is a formula $\varphi_{\mathbb{R}}$ such that

(2) $A \models \varphi_{\mathbb{R}}$ iff A and $\mathbb{R}$ have the same cardinality.

To obtain $\varphi_{\mathbb{R}}$, note that the ordered field $\mathfrak{R}^{<}$ of real numbers is, up to isomorphism, the only complete ordered field. Therefore, if ψ is the conjunction of the axioms for ordered fields and of the second-order $S_{\text{ar}}^{<}$-sentence

$$\forall X((\exists x Xx \wedge \exists y \forall z(Xz \rightarrow z < y)) \rightarrow \exists y(\forall z(Xz \rightarrow (z < y \vee z \equiv y)) \wedge \forall x(x < y \rightarrow \exists z(x < z \wedge Xz))))$$

("every nonempty set which is bounded above possesses a supremum"), then for all $S_{\text{ar}}^{<}$-structures $\mathfrak{A}$, the following holds:

(3) $\mathfrak{A} \models \psi$ iff $\mathfrak{A} \cong \mathfrak{R}^{<}$.

Hence, in order to obtain (2), we can choose as $\varphi_{\mathbb{R}}$ an $L_{\text{II}}^{\varnothing}$-sentence which says:

> "There are functions $+, \cdot$, elements 0, 1, and a relation $<$ such that ψ".

(We leave it to the reader to write down $\varphi_{\mathbb{R}}$ as a second-order sentence.) Now we can take as φ_{CH} a sentence which says that "if the domain is of the same cardinality as $\mathbb{R}$, then every subset of the domain is either at most countable or else of the same cardinality as the whole domain",

$$\varphi_{\text{CH}} := \varphi_{\mathbb{R}} \rightarrow \forall X(\varphi_{\leq \text{ctbl}}(X) \vee \exists g(\forall x Xgx \wedge \forall x \forall y(gx \equiv gy \rightarrow x \equiv y) \wedge \forall y(Xy \rightarrow \exists x gx \equiv y))).$$

It is easy to prove (cf. (1)) that

$$\models \varphi_{\text{CH}} \quad \text{iff CH holds.}$$

□

1.7 Exercise (The System $\mathscr{L}_{\text{II}}^{w}$ of *Weak Second-Order Logic*). For every S, let $L_{\text{II}}^{w,S} = L_{\text{II}}^{S}$. Change the notion of satisfaction for $\mathscr{L}_{\text{II}}$ by specifying, for $\mathfrak{I} = (\mathfrak{A}, \gamma)$ and n-ary X, that

$$\mathfrak{I} \models_{w} \exists X \varphi \quad \text{iff there is a } \textit{finite } C \subset A^{n} \text{ such that } \mathfrak{I}\frac{C}{X} \models_{w} \varphi.$$

Thus only quantification over finite sets (and relations) is allowed.

Show:

(a) There is a sentence φ and a structure $\mathfrak{A}$ such that $\mathfrak{A} \models_{w} \varphi$ but not $\mathfrak{A} \models \varphi$.
(b) For each sentence $\varphi \in L_{\text{II}}^{w,S}$, there is a sentence $\psi \in L_{\text{II}}^{S}$ such that for all S-structures $\mathfrak{A}$, $\mathfrak{A} \models_{w} \varphi$ iff $\mathfrak{A} \models \psi$.
(c) The compactness theorem does not hold for $\mathscr{L}_{\text{II}}^{w}$.

(However, the Löwenheim–Skolem theorem does hold for $\mathscr{L}_{\text{II}}^{w}$; cf. exercise 2.7 in the next section.)

§2. The System $\mathscr{L}_{\omega_1\omega}$

In VI.3.5 we showed that the class of torsion groups cannot be characterized in first-order logic. But we can axiomatize this class if we add to the group axioms the "formula"

$$(*) \qquad \forall x(x \equiv e \vee x \circ x \equiv e \vee x \circ x \circ x \equiv e \vee \cdots).$$

Thus we gain expressive power when allowing infinite disjunctions and conjunctions. Such formations are characteristic of the so-called *infinitary languages*. In the simplest case one restricts to conjunctions and disjunctions of countable length. This leads to the system $\mathscr{L}_{\omega_1\omega}$. (The notation $\mathscr{L}_{\omega_1\omega}$ follows the systematic terminology usual in the study of infinitary languages, cf. [7]).

To define the formulas of $\mathscr{L}_{\omega_1\omega}$ we use the jargon of calculi. Nevertheless it should be noted that the rule in 2.1(b) below is not a calculus rule in the strict sense, since it has infinitely many premises. (For example, in order to obtain the formula (*) one must already have obtained the formulas $x \equiv e$, $x \circ x \equiv e, \ldots$.) A precise version of such "calculi" and their usage can be given within the framework of set theory (cf. VII.4.3). For example, the definition of formulas and proofs by induction on formulas can be based on the principle of transfinite induction.

2.1 Definition of $\mathscr{L}_{\omega_1\omega}$. Compared with the first-order language L^S, we add the following to constitute the language $L^S_{\omega_1\omega}$:

(a) the symbol $\bigvee$ (for infinite disjunctions);
(b) to the calculus of formulas the following "rule":

> If Φ is an at most countable set of S-formulas, then $\bigvee\Phi$ is an S-formula (the *disjunction* of the formulas in Φ);

(c) to the definition of the notion of satisfaction the following clause:

> If Φ is an at most countable set of $L^S_{\omega_1\omega}$-formulas, $\mathfrak{A}$ an S-structure, β an assignment in $\mathfrak{A}$ and $\mathfrak{I} = (\mathfrak{A}, \beta)$ then
>
> $$\mathfrak{I} \models \bigvee\Phi \quad \text{iff} \quad \mathfrak{I} \models \varphi \text{ for some } \varphi \in \Phi.$$

There are many classes of structures which can be characterized in $L_{\omega_1\omega}$, but not in first-order logic. Examples are:

the class of torsion groups, characterized by the conjunction of the group axioms and

$$\forall x \bigvee\{\underbrace{x \circ \cdots \circ x}_{n\text{-times}} \equiv e \mid n \geq 1\},$$

the class of fields with characteristic a prime, by the conjunction of the field axioms and

$$\bigvee\{\underbrace{1 + \cdots + 1}_{n\text{-times}} \equiv 0 \mid n \text{ prime}\},$$

the class of archimedean ordered fields, by the conjunction of the axioms for ordered fields and

$$\forall x \bigvee\{x < \underbrace{1 + \cdots + 1}_{n\text{-times}} \mid n \geq 1\},$$

the class of structures isomorphic to $(\mathbb{N}, \sigma, 0)$, by the conjunction of the first two Peano axioms and

$$\forall x \bigvee\{x \equiv \underbrace{\sigma \cdots \sigma}_{n\text{-times}} 0 \mid n \geq 0\}.$$

2.2 Remarks.

(a) For a set Φ which is at most countable let $\bigwedge\Phi$ be an abbreviation for the $L_{\omega_1\omega}$-formula $\neg\bigvee\{\neg\varphi \mid \varphi \in \Phi\}$. Then

$$\mathfrak{I} \models \bigwedge\Phi \quad \text{iff for all } \varphi \in \Phi,\ \mathfrak{I} \models \varphi.$$

$\bigwedge\Phi$ is called the *conjunction* of the formulas in Φ.

(b) The definition of the set $\mathrm{SF}(\varphi)$ of subformulas of a formula φ in $L_{\omega_1\omega}$ is obtained from the corresponding definition for first-order formulas in II.4.5 by adding the clause $\mathrm{SF}(\bigvee\Phi) := \{\bigvee\Phi\} \cup \bigcup_{\psi\in\Phi} \mathrm{SF}(\psi)$. It can be proved for arbitrary φ that $\mathrm{SF}(\varphi)$ is at most countable. The proof is by induction on formulas; we give the $\bigvee$-step: Let $\varphi = \bigvee\Phi$, where by induction hypothesis $\mathrm{SF}(\psi)$ is at most countable for every $\psi \in \Phi$. Since $\mathrm{SF}(\varphi) = \{\varphi\} \cup \bigcup_{\psi\in\Phi} \mathrm{SF}(\psi)$ is an at most countable union of at most countable sets, $\mathrm{SF}(\varphi)$ is at most countable. In particular, for every $\varphi \in L^S_{\omega_1\omega}$ there exists an at most countable $S' \subset S$ such that $\varphi \in L^{S'}_{\omega_1\omega}$.

(c) Define the set $\mathrm{free}(\bigvee\Phi)$ of the variables occurring free in $\bigvee\Phi$ to be $\bigcup_{\psi\in\Phi} \mathrm{free}(\psi)$. The formula $\bigvee\{v_n \equiv v_n \mid n \in \mathbb{N}\}$ has infinitely many free variables. But one can easily prove by induction that in case $\mathrm{free}(\varphi)$ is finite then so is $\mathrm{free}(\psi)$ for any subformula ψ of φ. In particular, subformulas of an $\mathcal{L}_{\omega_1\omega}$-sentence have only finitely many free variables.

Consider the $L^{\varnothing}_{\omega_1\omega}$-sentence

$$\psi_{\text{fin}} := \bigvee\{\neg\varphi_{\geq n} \mid n \geq 2\}.$$

Then for every structure $\mathfrak{A}$ we have

$$\mathfrak{A} \models \psi_{\text{fin}} \quad \text{iff } A \text{ is finite.}$$

Hence the set of sentences $\{\psi_{\text{fin}}\} \cup \{\varphi_{\geq n} \mid n \geq 2\}$ is an example showing

2.3 Theorem. *The compactness theorem does not hold for $\mathcal{L}_{\omega_1\omega}$.* □

Nevertheless, many results for $\mathcal{L}_{\mathrm{I}}$ have their counterparts in $\mathcal{L}_{\omega_1\omega}$. We mention some examples and refer the reader to [20] for more information.

(1) The analogue of the Löwenheim–Skolem theorem holds (see 2.4 below).
(2) Extend the sequent calculus for first-order logic by the following "rules" for $\bigvee$:

$$(\bigvee\mathrm{A})\ \frac{\Gamma \quad \varphi \ \psi \ \text{for every } \varphi \in \Phi}{\Gamma \ \bigvee\Phi \ \psi}; \qquad (\bigvee\mathrm{S})\ \frac{\Gamma \quad \varphi}{\Gamma \ \bigvee\Phi}, \quad \text{if } \varphi \in \Phi,$$

where Γ stands for a finite sequence of $L_{\omega_1\omega}$-formulas. In this way one obtains a correct and complete "calculus": for $\mathscr{L}_{\omega_1\omega}$-sentences $\varphi_0, \varphi_1, \ldots, \varphi_{n-1}$ and φ, the sequent $\varphi_0, \varphi_1 \ldots \varphi_{n-1}\varphi$ is derivable if and only if it is correct. However, one must allow infinitely long derivations as is obvious from ($\bigvee$A).

(3) An analysis of (2) shows that by suitably generalizing the concept of finiteness one can transfer other results from $\mathscr{L}_{\mathrm{I}}$ to $\mathscr{L}_{\omega_1\omega}$. Among these is the *Barwise compactness theorem*, cf. [1].

2.4 Löwenheim–Skolem Theorem for $\mathscr{L}_{\omega_1\omega}$. *Every satisfiable $\mathscr{L}_{\omega_1\omega}$-sentence has a model over an at most countable domain.*

Since for every $\mathscr{L}_{\omega_1\omega}$-sentence φ there is an at most countable S such that $\varphi \in L^S_{\omega_1\omega}$, 2.4 follows directly from

2.5 Lemma. *Let S be at most countable, $\varphi \in L^S_{\omega_1\omega}$, and let $\mathfrak{B}$ be an S-structure satisfying φ. Then there is an at most countable substructure $\mathfrak{A} \subset \mathfrak{B}$ such that $\mathfrak{A} \models \varphi$.*

PROOF. We first present the idea of the proof.

Let B_0 be a nonempty at most countable subset of B which is S-closed, that is, which contains all $c^{\mathfrak{B}}$ for $c \in S$ and which is closed under application of $f^{\mathfrak{B}}$ for $f \in S$. Then B_0 is the domain of an at most countable substructure $\mathfrak{B}_0$ of $\mathfrak{B}$. If one tries to prove by induction that $\mathfrak{B}_0 \models \varphi$, the proof breaks down at the point where $\exists$-quantifiers are considered. For example, in a simple case where φ is of the form $\exists x Px$, one must ensure that there is a $b \in B_0$ such that $P^{\mathfrak{B}}b$. Therefore we shall close B_0 with respect to all possible existential requirements arising from subformulas of φ.

We now begin the proof. For pairwise distinct variables $x_0, \ldots, x_{n-1}$ we write $\psi(x_0, \ldots, x_{n-1})$ to denote a formula ψ with free$(\psi) \subset \{x_0, \ldots, x_{n-1}\}$. Define a sequence $A_0, A_1, A_2, \ldots$ of at most countable subsets of B so that

(a) $A_m \subset A_{m+1}$;
(b) for n-ary $f \in S$ and $a_0, \ldots, a_{n-1} \in A_m, f^{\mathfrak{B}}(a_0, \ldots, a_{n-1}) \in A_{m+1}$;
(c) for $\psi(x_0, \ldots, x_n) \in \mathrm{SF}(\varphi)$ and $a_0, \ldots, a_{n-1} \in A_m$, if

$$\mathfrak{B} \models \exists x_n \psi[a_0, \ldots, a_{n-1}]$$

then there is an $a_n \in A_{m+1}$ such that $\mathfrak{B} \models \psi[a_0, \ldots, a_{n-1}, a_n]$.

Let A_0 be a nonempty at most countable subset of B which contains $\{c^{\mathfrak{B}} \mid c \in S\}$. Suppose A_m is already defined and is at most countable. In order to define A_{m+1} we first set

$$A'_m := \{f^{\mathfrak{B}}(a_0, \ldots, a_{n-1}) \mid n \geq 1, f \in S \text{ } n\text{-ary}, a_0, \ldots, a_{n-1} \in A_m\}.$$

A'_m is also at most countable. Now, for $\psi(x_0, \ldots, x_n) \in \mathrm{SF}(\varphi)$ and $a_0, \ldots, a_{n-1} \in A_m$ such that $\mathfrak{B} \models \exists x_n \psi[a_0, \ldots, a_{n-1}]$, we choose an element $b \in B$

such that $\mathfrak{B} \models \psi[a_0, \ldots, a_{n-1}, b]$. Let A''_m be the set of b's chosen in this way. Since $\mathrm{SF}(\varphi)$ and A_m are at most countable, so is A''_m. Hence $A_{m+1} := A_m \cup A'_m \cup A''_m$ is also at most countable, and (a)–(c) are satisfied.

Now let

$$A := \bigcup_{m \in \mathbb{N}} A_m.$$

Then

(1) A is at most countable.
(2) A is S-closed. (By choice of A_0, we need only show that A is closed under the functions $f^{\mathfrak{B}}$. Let $f \in S$ be n-ary and $a_0, \ldots, a_{n-1} \in A$. Since the sets A_m form an ascending chain, $a_0, \ldots, a_{n-1}$ lie in some A_k. By (b) the element $f^{\mathfrak{B}}(a_0, \ldots, a_{n-1})$ lies in A_{k+1} and hence also in A.)

By (1) and (2), A is the domain of an at most countable substructure $\mathfrak{A}$ of $\mathfrak{B}$. Therefore we are done if we can show:

(*) $$\mathfrak{A} \models \varphi.$$

(*) follows immediately from the following claim:

(**) For all $\psi(x_0, \ldots, x_{n-1}) \in \mathrm{SF}(\varphi)$ and all $a_0, \ldots, a_{n-1} \in A$,

$$\mathfrak{A} \models \psi[a_0, \ldots, a_{n-1}] \quad \text{iff} \quad \mathfrak{B} \models \psi[a_0, \ldots, a_{n-1}].$$

We prove (**) by induction on ψ, but limit ourselves to the $\exists$-case.

Let $\psi(x_0, \ldots, x_{n-1}) = \exists x_n \chi(x_0, \ldots, x_n)$, and suppose $a_0, \ldots, a_{n-1} \in A$. If $\mathfrak{A} \models \exists x_n \chi[a_0, \ldots, a_{n-1}]$ then we obtain successively:

There is $a \in A$ such that $\mathfrak{A} \models \chi[a_0, \ldots, a_{n-1}, a]$.

There is $a \in A$ such that $\mathfrak{B} \models \chi[a_0, \ldots, a_{n-1}, a]$ (Ind. hyp.).

$\mathfrak{B} \models \exists x_n \chi[a_0, \ldots, a_{n-1}]$.

Conversely, if $\mathfrak{B} \models \exists x_n \chi[a_0, \ldots, a_{n-1}]$, we choose k such that $a_0, \ldots, a_{n-1} \in A_k$, and we obtain successively:

There is $a \in A_{k+1}$ such that $\mathfrak{B} \models \chi[a_0, \ldots, a_{n-1}, a]$ (by (c)).

There is $a \in A_{k+1}$ such that $\mathfrak{A} \models \chi[a_0, \ldots, a_{n-1}, a]$ (Ind. hyp.).

$\mathfrak{A} \models \exists x_n \chi[a_0, \ldots, a_{n-1}]$. □

Consider an at most countable set Φ of first-order sentences and let $\varphi := \bigwedge \Phi$. Then it follows from 2.5 that every model of Φ has an at most countable substructure which is also a model of Φ. In particular, this yields a proof of the Löwenheim–Skolem theorem for first-order logic which does not rely on the proof of the completeness theorem. Note that an $\mathcal{L}_{\omega_1\omega}$-sentence characterizing $(\mathbb{N}, \sigma, 0)$ has no uncountable model; hence in $\mathcal{L}_{\omega_1\omega}$ we do not have an analogue of the upward Löwenheim–Skolem theorem VI.2.3.

To conclude this section we give a mathematical application of 2.5 by choosing φ appropriately.

We consider groups as S-structures with $S = \{\circ, e, ^{-1}\}$. A group $\mathfrak{G}$ is said to be *simple* if $\{e^G\}$ and G are the only normal subgroups of $\mathfrak{G}$. If for $a \in G$ we denote by $\langle a \rangle^{\mathfrak{G}}$ the normal subgroup of $\mathfrak{G}$ generated by a, then clearly $\mathfrak{G}$ is simple if and only if $\langle a \rangle^{\mathfrak{G}} = G$ for all $a \in G$ such that $a \neq e^G$.

Since

$$\langle a \rangle_{\mathfrak{G}} = \{g_0 a^{z_0} g_0^{-1} \dots g_n a^{z_n} g_n^{-1} \mid n \in \mathbb{N}, z_0, \dots, z_n \in \mathbb{Z}, g_0, \dots, g_n \in G\},$$

the class of simple groups can be axiomatized in $L^S_{\omega_1\omega}$ by the conjunction φ_0 of the group axioms and the following sentence:

$$\forall x(\neg x \equiv e \rightarrow \forall y \bigvee\{\exists v_0 \dots v_n \bigvee\{y \equiv v_0 x^{z_0} v_0^{-1} \dots v_n x^{z_n} v_n^{-1} \mid z_0, \dots, z_n \in \mathbb{Z}\} \mid n \in \mathbb{N}\}.$$

Using 2.5 we now show

2.6 Proposition. *If $\mathfrak{G}$ is a simple group and M a countable subset of G then there is a countable simple subgroup of $\mathfrak{G}$ which contains M.*

Proof. Let $\bar{S} := S \cup \{c_a \mid a \in M\}$, where the c_a are new constants for $a \in M$. We expand $\mathfrak{G}$ to an $\bar{S}$-structure $\bar{\mathfrak{G}}$, interpreting each c_a by the corresponding a, and apply 2.5 to $\bar{\mathfrak{G}}$ and φ_0. □

2.7 Exercise. Show that for every $L^{w,S}_{\mathrm{II}}$-sentence φ (cf. exercise 1.7) there is an $L^S_{\omega_1\omega}$-sentence ψ with the same models (that is, for all S-structures $\mathfrak{A}$, $\mathfrak{A} \models_w \varphi$ iff $\mathfrak{A} \models \psi$). Conclude from this that the Löwenheim–Skolem theorem holds for $\mathscr{L}^w_{\mathrm{II}}$.

2.8 Exercise. Show that the following classes can be axiomatized by an $\mathscr{L}_{\omega_1\omega}$-sentence:

(a) the class of finitely generated groups;
(b) the class of structures isomorphic to $(\mathbb{Z}, <)$.

2.9 Exercise. (a) For arbitrary S, show that $L^S_{\omega_1\omega}$ is uncountable.
(b) Give an uncountable structure $\mathfrak{B}$ (for a suitable countable symbol set S) such that there is no countable structure $\mathfrak{A}$ satisfying the same $L^S_{\omega_1\omega}$-sentences.

2.10 Exercise. Using 2.5, show that any two infinite $\varnothing$-structures satisfy the same $\mathscr{L}^{\varnothing}_{\omega_1\omega}$-sentences.

§3. The System $\mathscr{L}_Q$

The system $\mathscr{L}_Q$ is obtained from first-order logic by adding the quantifier Q, where a formula $Qx\varphi$ says "there are uncountably many x satisfying φ".

3.1 Definition of $\mathscr{L}_Q$. Compared with the first-order language L^S, we add the following to constitute the language L_Q^S:

(a) the symbol Q;
(b) to the calculus of formulas, the rule:

$$\text{If } \varphi \text{ is an } S\text{-formula then so is } Qx\varphi;$$

(c) to the definition of the notion of satisfaction, the clause:
If φ is an S-formula and $\mathfrak{I} = (\mathfrak{A}, \beta)$ an S-interpretation then

$$\mathfrak{I} \models Qx\varphi \quad \text{iff} \quad \left\{a \in A \mid \mathfrak{I}\frac{a}{x} \models \varphi\right\} \text{ is uncountable.}$$

$\mathscr{L}_Q$ has more expressive power than $\mathscr{L}_I$. For example, the class of at most countable structures can be axiomatized in $\mathscr{L}_Q$ by the sentence $\neg Qx\, x \equiv x$. For $S = \{<\}$ let φ_0 be the conjunction of the axioms for orderings and $(Qx\, x \equiv x \wedge \forall x \neg Qy\, y < x)$. Then φ_0 is an L_Q^S-sentence characterizing the class of uncountable orderings in which every element has at most countably many predecessors. These so-called ω_1-like orderings play an important rôle in investigations of $\mathscr{L}_Q$.

Note that the sentence φ_0, or even the sentence $Qx\, x \equiv x$, has an uncountable, but no at most countable model. Hence the strict analogue of the Löwenheim–Skolem theorem does not hold. (However, each satisfiable $\mathscr{L}_Q$-sentence has a model of cardinality $\leq \aleph_1$, cf. exercise 3.3.)

One can set up an adequate sequent calculus $\mathfrak{S}_Q$ for $\mathscr{L}_Q$ by adding the following rules to the sequent calculus for first-order logic. (After each rule an explanatory comment is given, which is also the essence of a correctness proof.)

$$\frac{\Gamma\ Qx\varphi}{\Gamma\ Qy\varphi\frac{y}{x}}, \quad \text{if } y \text{ is not free in } \varphi.$$

(Renaming of bound variables);

$$\frac{}{\neg\ Qx(x \equiv y \vee x \equiv z)}, \quad \text{if } y \text{ and } z \text{ are distinct from } x.$$

("Singletons and pair sets are not uncountable");

$$\frac{\Gamma\ \forall x(\varphi \rightarrow \psi)}{\Gamma\ Qx\varphi \rightarrow Qx\psi}.$$

("Sets having uncountable subsets are uncountable");

$$\begin{array}{ll} \Gamma & \neg Qx\exists y\varphi \\ \Gamma & Qy\exists x\varphi \\ \hline \Gamma & \exists xQy\varphi \end{array}.$$

("If the union of at most countably many sets is uncountable then at least one of these sets is uncountable").

One can show (cf. [19]) that the calculus $\mathfrak{S}_Q$ is correct and complete in the sense that the equivalence "$\Phi \models \varphi$ iff $\Phi \vdash \varphi$" holds for $\mathscr{L}_Q$ if Φ is at most countable. As for first-order logic we conclude (cf. VI.2):

3.2 $\mathscr{L}_Q$-Compactness Theorem. *For every countable set Φ of L_Q^S-formulas, Φ is satisfiable if and only if every finite subset of Φ is satisfiable.* □

The following example shows that the compactness theorem does not hold for uncountable sets of formulas. Let S be an uncountable set of constants and let

$$\Phi := \{\neg c \equiv d \mid c, d \in S, c \neq d\} \cup \{\neg Qx\, x \equiv x\}.$$

Then every finite subset of Φ is satisfiable, but Φ itself is not.

In Chapter VI we saw that the compactness theorem and the Löwenheim–Skolem theorem are useful for mathematical applications. None of the extensions of $\mathscr{L}_\mathrm{I}$ which we have discussed in this section satisfies both theorems. The compactness theorem fails for $\mathscr{L}_{\omega_1\omega}$, the Löwenheim–Skolem theorem for $\mathscr{L}_Q$, and both for $\mathscr{L}_\mathrm{II}$. Does there exist any logical system at all which has more expressive power than first-order logic and for which both the compactness theorem and the Löwenheim–Skolem theorem hold? We shall give a negative answer to this question in Chapter XII.

3.3 Exercise. Show that every satisfiable $\mathscr{L}_Q$-sentence has a model over a domain of cardinality at most $\aleph_1$ (where $\aleph_1$ is the smallest uncountable cardinal). (*Hint:* Use a method similar to that in the proof of 2.5: for formulas $Qx\varphi$, which hold in $\mathfrak{B}$, add $\aleph_1$ elements satisfying φ.)

3.4 Exercise. Let $\mathscr{L}_Q^0$ be obtained from $\mathscr{L}_Q$ by changing the notion of satisfaction 3.1(c) as follows:

$$\mathfrak{I} \models Qx\varphi \quad \text{iff} \quad \left\{a \in A \,\middle|\, \mathfrak{I}\frac{a}{x} \models \varphi\right\} \text{ is at most countable.}$$

Show that the compactness theorem does not hold for $\mathscr{L}_Q^0$, but that the Löwenheim–Skolem theorem does.

CHAPTER X

Limitations of the Formal Method

Only in methodological questions have we thus far referred to the fact that applications of sequent rules consist ultimately of mechanical operations on symbol strings (cf. VII.1). In the following we also want to make stronger use of this formal-syntactic aspect in mathematical considerations. Let us give an initial idea, taking as an example the system of axioms $\Phi_{gr} = \{\varphi_0, \varphi_1, \varphi_2\}$ for group theory. It follows from the completeness theorem that for all S_{gr}-sentences φ,

$$\Phi_{gr} \models \varphi \quad \text{iff } \Phi_{gr} \vdash \varphi.$$

Thus φ is a theorem of group theory

$$\mathrm{Th}_{gr} = \{\psi \in L_0^{S_{gr}} \mid \Phi_{gr} \models \psi\},$$

if and only if the sequent $\varphi_0 \varphi_1 \varphi_2 \varphi$ is derivable.

By systematically applying all the sequent rules one can generate all possible derivations and thus compile a list of the theorems of Th_{gr}: One adds a sentence $\varphi \in L_0^{S_{gr}}$ to the list if one arrives at a derivation whose last sequent is $\varphi_0 \varphi_1 \varphi_2 \varphi$. Hence there is a procedure by which one can in a "mechanical" way list all theorems of Th_{gr}. It should be plausible that one could use a suitably programmed computer to carry out such a procedure. Of course, one would have to be able to increase the capacity of the computer if necessary since the derivations and the sequents and formulas therein can be arbitrarily long. A set such as Th_{gr} which can be listed by means of such a procedure is said to be *enumerable.*

Of course, the above enumeration procedure yields many trivialities like $\forall x(x \equiv x \rightarrow x \equiv x)$. On the other hand, a group theorist is only interested in specific group-theoretical statements φ which are relevant for his investigations. His aim is to determine for such a φ whether $\varphi \in \mathrm{Th}_{gr}$ or not.

Usually this is accomplished either by a proof or by a counterexample. Unfortunately, an enumeration procedure for $\mathrm{Th}_{\mathrm{gr}}$ as above is of little help in this situation: Given φ, one might start the procedure in order to see whether φ appears as an output; however, if $\varphi \notin \mathrm{Th}_{\mathrm{gr}}$ the procedure will not yield this information since at any step one is left uncertain as to whether φ will appear later or will not appear at all. Thus we are led to seek a different kind of procedure which can be applied to an arbitrary S_{gr}-sentence φ and then stops after finitely many steps, yielding the decision whether $\varphi \in \mathrm{Th}_{\mathrm{gr}}$ or not. Put another way, can one program a computer so that whenever it is given an S_{gr}-sentence φ it "computes" whether φ belongs to $\mathrm{Th}_{\mathrm{gr}}$? If such a procedure exists for a given theory, we call that theory *decidable*.

The present chapter is devoted to questions of this kind. First we discuss the concepts of enumerability and decidability in more detail, in §1 from a naive point of view, and in §2 on the basis of the precise notion of *register machine*. These topics form part of the so-called *recursion theory* (*theory of computability*). The remaining sections of the chapter then contain applications to first-order and second-order logic.

For further information about recursion theory we refer the reader to [6] and [15].

§1. Decidability and Enumerability

A. Procedures, Decidability

It is well known how to decide whether an arbitrary natural number n is prime: If $n = 0$ or $n = 1$, n is not prime. If $n \geq 2$, one tests the numbers $2, 3, \ldots, n - 1$ to see whether they divide n. If none of these numbers divides n then n is prime; otherwise it is not.

This procedure operates with strings of symbols. For example, in the case of decimal representation of natural numbers it operates with strings over the alphabet $\{0, \ldots, 9\}$. Our description has not specified it in complete detail—for instance, we have not described how division is to be carried out—but it should be clear that it is possible to fill these gaps in order to ensure that all steps are completely determined. In view of its purpose we call the procedure a *decision procedure for the set of primes*.

Other procedures which are well known include those for

(a) multiplying two natural numbers,
(b) computing the square root of a natural number,
(c) listing the primes in increasing order.

Common to all of these procedures is the fact that they proceed step by step, they operate on symbol strings of a well-defined sort, and they can be carried out by a suitably programmed computer. A procedure can operate on one or more *inputs* (as in (a) or (b)) or it can be started without any particular input (as in (c)). It can *stop* after finitely many steps and yield an *output* (as in (a) for any input and in (b) for inputs which are squares), or it can run without ever stopping, possibly giving an output from time to time (as in (c)).

Procedures in our sense (effective procedures, processes, algorithms) operate with concrete objects such as symbol strings. Sometimes mathematicians use these notions in a wider sense, speaking, for instance, of the Gram–Schmidt orthogonalization process even when referring to abstract vector spaces.

Concerning the following definition and the subsequent discussion the reader should bear in mind that the notion of procedure has so far been introduced just in an intuitive way and by means of examples.

1.1 Definition. Let $\mathbb{A}$ be an alphabet, W a set of words over $\mathbb{A}$, i.e., $W \subset \mathbb{A}^*$, and $\mathfrak{P}$ a procedure.

(a) $\mathfrak{P}$ is a *decision procedure* for W if, for every input $\zeta \in \mathbb{A}^*$, $\mathfrak{P}$ eventually stops, having previously given exactly one output $\eta \in \mathbb{A}^*$, where

$$\eta = \square, \quad \text{if } \zeta \in W,$$
$$\eta \neq \square, \quad \text{if } \zeta \notin W.$$

(b) W is *decidable* if there is a decision procedure for W.

Thus when a decision procedure for W is applied to an arbitrary word ζ over $\mathbb{A}$, it yields an answer to the question "$\zeta \in W$?" in finitely many steps. The answer is "yes" if the output is the empty word; it is "no" if the output is a nonempty word.

To formulate the above decision procedure for the set of primes according to definition 1.1 we set $\mathbb{A} := \{0, \ldots, 9\}$ and $W :=$ the set of primes, and we agree that the empty word shall be the output for primes and, say, 1 the output for nonprimes.

Further examples of decidable sets are the set of terms and the set of formulas for a concretely given symbol set. In the case of S_∞ (cf. II.2) for instance, terms and formulas are strings over the alphabet

$$\mathbb{A}_\infty := \{v_0, v_1, \ldots, \neg, \vee, \exists, \equiv,), (\} \cup S_\infty.$$

We sketch a decision procedure for the terms.

Let $\zeta \in \mathbb{A}_\infty^*$ be given. First determine the length $l(\zeta)$. If $l(\zeta) = 0$, ζ is not a term. If $l(\zeta) = 1$, ζ is a term if and only if ζ is a variable or a constant. If $l(\zeta) > 1$, ζ is not a term unless it begins with a function symbol. If ζ begins with a function symbol, say $\zeta = f_1^3\zeta'$, then check whether there is a decomposition $\zeta' = \zeta_0\zeta_1\zeta_2$, where the ζ_i are terms. ζ is a term if and only if such a decomposition exists. To check whether each ζ_i is a term, use the same procedure as for ζ. (Clearly, in this way an answer will be obtained after finitely many steps.)

If one analyzes the procedure or tries to program it for a computer, a difficulty arises: programs (or descriptions of procedures) are finite and therefore can only refer to finitely many symbols in $\mathbb{A}_\infty$, whereas $\mathbb{A}_\infty$ contains, among other things, the infinite list of symbols $v_0, v_1, v_2, \ldots$. Therefore we introduce the new finite alphabet

$$\mathbb{A}_0 := \{v, \underline{0}, \underline{1}, \ldots, \underline{9}, \bar{0}, \bar{1}, \ldots, \bar{9}, \neg, \vee, \exists, \equiv,), (, R, f, c\}$$

and then represent the symbols in $\mathbb{A}_\infty$ in terms of the symbols of $\mathbb{A}_0$ in the natural way. For example, we represent v_{71} by $v\underline{7}\underline{1}$, c_{11} by $c\underline{1}\underline{1}$, R^3_{18} by $R\bar{3}\underline{1}\underline{8}$ and the S_∞-formula $\exists v_3(R^1_1 v_3 \vee c_{11} \equiv f^1_0 v_1)$ by $\exists v\underline{3}(R\bar{1}\underline{1}v\underline{3} \vee c\underline{1}\underline{1} \equiv f\bar{1}\underline{0}v\underline{1})$. With this in mind we only consider *finite* alphabets in the sequel.

1.2 Exercise. Let $\mathbb{A}$ be an alphabet, and let W, W' be decidable subsets of $\mathbb{A}^*$. Show that $W \cup W'$, $W \cap W'$, and $\mathbb{A}^* - W$ are also decidable.

1.3 Exercise. Describe decision procedures for the following subsets of $\mathbb{A}_0^*$:

(a) the set of strings $x\varphi$ over $\mathbb{A}_0$ such that $x \in \text{free}(\varphi)$,
(b) the set of S_∞-sentences.

B. Enumerability

We consider a computing machine which operates as follows: it successively generates the numbers 0, 1, 2, ..., tests in each case whether n is a prime, and yields n as output if the answer is positive. The machine runs without ever stopping, and it generates a list of all primes, i.e., a list in which every prime eventually appears.

Sets, such as the primes, which can be listed by means of a procedure are said to be *enumerable*:

1.4 Definition. Let $\mathbb{A}$ be an alphabet, $W \subset \mathbb{A}^*$ and $\mathfrak{P}$ a procedure.

(a) $\mathfrak{P}$ is an *enumeration procedure* for W if $\mathfrak{P}$, once having been started, eventually yields as outputs exactly the words in W (in some order, possibly with repetitions).
(b) W is *enumerable* if there is an enumeration procedure for W.

We give some further examples for enumerable sets.

1.5 Proposition. *If $\mathbb{A}$ is a (finite) alphabet, then $\mathbb{A}^*$ is enumerable.*

Proof. Suppose $\mathbb{A} = \{a_0, \ldots, a_n\}$. We first define the *lexicographic order* on $\mathbb{A}^*$ (with respect to the indexing $a_0, \ldots, a_n$). In this ordering ζ precedes ζ' if either

$$l(\zeta) < l(\zeta')$$

or

$$l(\zeta) = l(\zeta') \quad \text{and} \quad \text{“}\zeta \text{ precedes } \zeta' \text{ in a dictionary”},$$

that is, there are $a_i, a_j \in \mathbb{A}$, with $i < j$, such that for suitable $\xi, \eta, \eta' \in \mathbb{A}^*$, $\zeta = \xi a_i \eta$ and $\zeta' = \xi a_j \eta'$.

For example, if $\mathbb{A} = \{a, b, c, \ldots, x, y, z\}$, then “papa” comes before “papi”, but after “zoo”. In general the ordering begins as follows:

$$\square, a_0, \ldots, a_n, a_0 a_0, a_0 a_1, \ldots, a_0 a_n, a_1 a_0, \ldots, a_n a_n, a_0 a_0 a_0, \ldots .$$

It is easy to set up a procedure which lists the elements of $\mathbb{A}^*$ in lexicographic order. $\square$

1.6 Proposition. *$\{\varphi \in L_0^{S_\infty} \mid \models \varphi\}$ is enumerable.*

Proof. By the completeness theorem we have to describe a procedure which lists the S_∞-sentences φ with $\vdash \varphi$. We use the same idea as in the procedure for listing $\mathrm{Th}_{\mathrm{gr}}$ at the beginning of this chapter: We systematically generate all possible derivations for the symbol set S_∞. If the last sequent in such a derivation consists of a single sentence φ, we include φ in the list. Note that the derivations can be generated as follows: For $n = 1, 2, 3, \ldots$ one constructs the first n terms and formulas in the lexicographical ordering, and one forms the finitely many derivations of length $\leq n$ which use only these formulas and terms, and which consist of sequents containing at most n members. $\square$

C. The Relationship Between Enumerability and Decidability

We have just seen that the set of “logically true” sentences can be listed by means of an enumeration procedure. Is it possible to go farther than this and *decide* whether an arbitrary given sentence is “logically true”? The enumeration procedure given above does not help to solve this problem. For example, if we want to test a sentence φ_0 for validity we might start the enumeration procedure in 1.6 and wait to see whether φ_0 appears; we obtain a positive decision as soon as φ_0 is added to the list. But as long as φ_0 has not appeared we cannot say anything about φ_0 since we do not know whether φ_0 will never appear (because it is not valid) or whether it will appear at a later time. In fact we shall show (cf. 4.1) that the set of valid S_∞-sentences is not decidable.

On the other hand, if a set is decidable we can conclude that it is enumerable:

1.7 Theorem. *Every decidable set is enumerable.*

PROOF. Suppose $W \subset \mathbb{A}^*$ is decidable and $\mathfrak{P}$ is a decision procedure for W. To list W, generate the strings of $\mathbb{A}^*$ in lexicographic order, use $\mathfrak{P}$ to check for each string ζ whether it belongs to W or not, and, if the answer is positive, add ζ to the list. □

As an extension of 1.7 we have:

1.8 Theorem. *A subset W of $\mathbb{A}^*$ is decidable if and only if W and the complement $\mathbb{A}^* - W$ are enumerable.*

PROOF. Suppose W is decidable. Then $\mathbb{A}^* - W$ is also decidable (one can use a decision procedure for W, merely interchanging the outputs "yes" and "no"). Thus by 1.7, W and $\mathbb{A}^* - W$ are enumerable. Conversely, suppose W and $\mathbb{A}^* - W$ are enumerable by means of procedures $\mathfrak{P}$ and $\mathfrak{P}'$. We combine $\mathfrak{P}$ and $\mathfrak{P}'$ into a decision procedure for W which operates as follows: Given ζ, $\mathfrak{P}$ and $\mathfrak{P}'$ run simultaneously until ζ is yielded by either $\mathfrak{P}$ or $\mathfrak{P}'$. This will eventually be the case since every symbol string in $\mathbb{A}^*$ is either in W or in $\mathbb{A}^* - W$. If ζ is listed by $\mathfrak{P}$, the output is "yes" ($\zeta \in W$), if it is listed by $\mathfrak{P}'$, the output is "no" ($\zeta \notin W$). □

1.9 Exercise. Suppose $U \subset \mathbb{A}^*$ is decidable and $W \subset U$. Show that if W and $U - W$ are enumerable, then W is decidable.

Our definitions of decidability and enumerability were given with respect to a fixed alphabet. However, this reference is not essential:

1.10 Exercise. Let $\mathbb{A}_1$ and $\mathbb{A}_2$ be alphabets such that $\mathbb{A}_1 \subset \mathbb{A}_2$, and suppose $W \subset \mathbb{A}_1^*$. Show that W is decidable (enumerable) with respect to $\mathbb{A}_1$ if and only if it is decidable (enumerable) with respect to $\mathbb{A}_2$.

D. Computable Functions

Let $\mathbb{A}$ and $\mathbb{B}$ be alphabets. A procedure which for each input from $\mathbb{A}^*$ yields a word in $\mathbb{B}^*$ determines a function from $\mathbb{A}^*$ to $\mathbb{B}^*$. A function whose values can be computed in this way by a procedure is said to be *computable*. An example of a computable function is the length function l, which assigns to every $\zeta \in \mathbb{A}^*$ the length of ζ (in decimal notation as a word over the alphabet $\{0, \ldots, 9\}$).

Whereas our discussion of effective procedures deals mainly with the notions of enumerability and decidability, many presentations of recursion theory start with computability of functions as the key concept. Both approaches are equivalent in the sense that the above notions are definable from each other. The following exercise shows that the notion of computable function can be reduced to both the notion of enumerability and the notion of decidability.

1.11 Exercise. Let $\mathbb{A}$, $\mathbb{B}$ be alphabets, $\# \notin \mathbb{A} \cup \mathbb{B}$, and $f: \mathbb{A}^* \to \mathbb{B}^*$. Show that the following are equivalent:

(i) f is computable.
(ii) $\{w \# f(w) \mid w \in \mathbb{A}^*\}$ is enumerable.
(iii) $\{w \# f(w) \mid w \in \mathbb{A}^*\}$ is decidable.

$\{w \# f(w) \mid w \in \mathbb{A}^*\}$ can be considered as the graph of f, and hence the equivalences in 1.11 can be formulated as follows. A function is computable iff its graph is enumerable (decidable). Note that by 1.8 the notion of decidability can be reduced to that of enumerability.

§2. Register Machines

In the foregoing discussion we have used an intuitive notion of procedure which we illustrated by use of examples. The conception we have thus acquired is perhaps sufficient for recognizing in a given case whether a proposed procedure can be accepted as such. But in general, our informal concept does not enable us to prove that a particular set is not decidable. Namely, in this case one must show that *every* possible procedure is not a decision procedure for the set in question. But such a proof is usually not possible without a precise notion of procedure.

We now introduce such a precise concept, starting from the idea that a procedure should be programmable on a computer. For this purpose we set up a programming language and define procedures in the formal sense to be exactly those procedures which can be programmed in this language.

For the following discussion we fix an alphabet

$$\mathbb{A} = \{a_0, \ldots, a_r\}.$$

The programs are executed by computers with a memory consisting of units $R_0, \ldots, R_m$, called *registers*. (In the literature such machines are frequently called *register machines*.) At each stage in a computation every register contains a word from $\mathbb{A}^*$. We assume that we have machines with arbitrarily many registers at our disposal, and that the individual registers can store words of arbitrary length. This idealization agrees with our objective of encompassing all procedures which can be carried out in principle by a computer, i.e., disregarding problems of capacity.

A program (over $\mathbb{A} = \{a_0, \ldots, a_r\}$) consists of instructions, where each instruction begins with a natural number L, its *label*. Only instructions of the form (1) through (5) below are permitted. (The instructions describe very elementary operations on registers; for instance, they will add or delete a symbol in a register, or will test whether a register contains the empty word or not. Together with the instructions we give their precise meaning in parentheses.)

(1) $\mathrm{L\ LET\ R_i = R_i + a_j}$
for $L, i, j \in \mathbb{N}$ with $j \leq r$ (Add-instruction: "Add the letter a_j to the word in register R_i");

(2) $\mathrm{L\ LET\ R_i = R_i - a_j}$
for $L, i, j \in \mathbb{N}$ with $j \leq r$ (Subtract-instruction: "If the word in register R_i ends with the letter a_j, delete this a_j; otherwise leave the word unchanged");

(3) $\mathrm{L\ IF\ R_i = \square\ THEN\ L'\ ELSE\ L_0\ OR \ldots OR\ L_r}$
for $L, i, L', L_0, \ldots, L_r \in \mathbb{N}$ (Jump-instruction: "If register R_i contains the empty word go to instruction labelled L'; if the word in register R_i ends with a_0 (resp. $a_1, \ldots, a_r$) go to instruction labelled L_0 (resp. $L_1, \ldots, L_r$)");

(4) L PRINT
for $L \in \mathbb{N}$ (Print-instruction: "Print as output the word stored in register R_0");

(5) L HALT
for $L \in \mathbb{N}$ (Halt-instruction: "Halt").

2.1 Definition. A *register program* (or simply *program*) is a finite sequence $\alpha_0, \ldots, \alpha_k$ of instructions of the form (1) through (5) with the following properties:

(i) α_i has label i ($i = 0, \ldots, k$).
(ii) In an instruction of the form (3) the labels $L', L_0, \ldots, L_r$ are $\leq k$.
(iii) Only α_k is a halt-instruction.

Each program P gives rise to a procedure: Imagine we have a computer which contains all registers occurring in P and which has been programmed with P. At the beginning of a computation all registers with the possible exception of R_0 are empty, i.e., they contain the empty word, whereas R_0 contains a possible input. The computation proceeds stepwise, each step corresponding to the execution of one instruction of the program. Beginning with the first instruction one proceeds line by line through the program, jumping only as required by a jump-instruction. Whenever a print-instruction is encountered, the respective content of R_0 is given as an output ("printed out"). The machine stops when the halt-instruction is reached. Some examples of programs follow.

2.2 Example. Let $\mathbb{A} = \{|\}$. We interpret the strings $\square, |, \|, \ldots$ as the natural numbers $0, 1, 2, \ldots$. The following program P_0 decides whether an input in the register R_0 is an even number or not: P_0 successively deletes strokes from the string n given as an input in R_0 until the empty string is obtained. It ascertains whether n is even or odd and prints out $\square$ or $|$ accordingly and then stops.

0 IF $R_0 = \square$ THEN 6 ELSE 1
1 LET $R_0 = R_0 - |$
2 IF $R_0 = \square$ THEN 5 ELSE 3
3 LET $R_0 = R_0 - |$
4 IF $R_0 = \square$ THEN 6 ELSE 1
5 LET $R_0 = R_0 + |$
6 PRINT
7 HALT.

We say that a program P is *started* with a word $\zeta \in \mathbb{A}^*$, if P begins the computation with ζ in R_0 and $\square$ in the remaining registers. If P, started with ζ, eventually reaches the halt-instruction, we write

$$P: \zeta \to \text{halt};$$

otherwise we write

$$P: \zeta \to \infty.$$

For $\zeta, \eta \in \mathbb{A}^*$,

$$P: \zeta \to \eta$$

means that P started with ζ eventually stops, having—in the course of the computation—given exactly one output, namely η. In the above example,

$$P_0: n \to \square, \quad \text{if } n \text{ is even,}$$

$$P_0: n \to |, \quad \text{if } n \text{ is odd.}$$

2.3 Example. Let $\mathbb{A} = \{a_0, \ldots, a_r\}$. For the program P:

0 PRINT
1 LET $R_0 = R_0 + a_0$
2 IF $R_0 = \square$ THEN 0 ELSE 0 OR ... OR 0
3 HALT

we have $P: \zeta \to \infty$ for all ζ. If P is started with a word ζ, P prints out successively the words $\zeta, \zeta a_0, \zeta a_0 a_0, \ldots$.

Line 2 of P has the form

$$\text{L IF } R_0 = \square \text{ THEN L}' \text{ ELSE L}' \text{ OR} \ldots \text{OR L}'.$$

In every case such an instruction results in a jump to line L'. For the sake of simplicity we shall in the sequel abbreviate it by

$$\text{L GOTO L}'.$$

2.4 Example. We present a program P for the alphabet $\mathbb{A} = \{a_0, a_1\}$, such that $P: \zeta \to \zeta\zeta$ for $\zeta \in \mathbb{A}^*$. (Given ζ in R_0, instructions 0–8 serve to build up ζ in reverse order in the registers R_1 and R_2, after which $\zeta\zeta$ is built up in R_0, taking the first copy from R_1 (instructions 9–15) and the second copy from R_2 (instructions 16–22).)

0 IF $\mathrm{R}_0 = \square$ THEN 9 ELSE 1 OR 5
1 LET $\mathrm{R}_0 = \mathrm{R}_0 - \mathrm{a}_0$
2 LET $\mathrm{R}_1 = \mathrm{R}_1 + \mathrm{a}_0$
3 LET $\mathrm{R}_2 = \mathrm{R}_2 + \mathrm{a}_0$
4 GOTO 0
5 LET $\mathrm{R}_0 = \mathrm{R}_0 - \mathrm{a}_1$
6 LET $\mathrm{R}_1 = \mathrm{R}_1 + \mathrm{a}_1$
7 LET $\mathrm{R}_2 = \mathrm{R}_2 + \mathrm{a}_1$
8 GOTO 0
9 IF $\mathrm{R}_1 = \square$ THEN 16 ELSE 10 OR 13
10 LET $\mathrm{R}_1 = \mathrm{R}_1 - \mathrm{a}_0$
11 LET $\mathrm{R}_0 = \mathrm{R}_0 + \mathrm{a}_0$
12 GOTO 9
13 LET $\mathrm{R}_1 = \mathrm{R}_1 - \mathrm{a}_1$
14 LET $\mathrm{R}_0 = \mathrm{R}_0 + \mathrm{a}_1$
15 GOTO 9
16 IF $\mathrm{R}_2 = \square$ THEN 23 ELSE 17 OR 20
17 LET $\mathrm{R}_2 = \mathrm{R}_2 - \mathrm{a}_0$
18 LET $\mathrm{R}_0 = \mathrm{R}_0 + \mathrm{a}_0$
19 GOTO 16
20 LET $\mathrm{R}_2 = \mathrm{R}_2 - \mathrm{a}_1$
21 LET $\mathrm{R}_0 = \mathrm{R}_0 + \mathrm{a}_1$
22 GOTO 16
23 HALT.

As an exercise the reader should write a program P over the alphabet $\mathbb{A} = \{a_0, a_1, a_2\}$ which accomplishes the following:

$$P: \zeta \to \text{halt}, \quad \text{if } \zeta = a_0 a_0 a_2,$$

$$P: \zeta \to \infty, \quad \text{if } \zeta \neq a_0 a_0 a_2.$$

By analogy with the naive definitions in §1, we can introduce the exact notions of register-decidability and register-enumerability.

2.5 Definition. Suppose $W \subset \mathbb{A}^*$.

(a) A program P *decides* W if for all $\zeta \in \mathbb{A}^*$,

$$P: \zeta \to \square, \qquad \text{if } \zeta \in W,$$

$$P: \zeta \to \eta \text{ with } \eta \neq \square, \quad \text{if } \zeta \notin W.$$

(b) W is said to be *register-decidable* (abbreviated: *R-decidable*) if there is a program which decides W.

Example 2.2 shows that the set of even natural numbers is R-decidable.

2.6 Definition. Let $W \subset \mathbb{A}^*$.

(a) A program P *enumerates* W if P, started with $\square$, prints out exactly the words in W (in any order, possibly with repetitions).
(b) W is said to be *register-enumerable* (abbreviated: *R-enumerable*), if there is a program which enumerates W.

If P enumerates an infinite set, then $P: \square \to \infty$. By 2.3, $W = \{\square, a_0, a_0 a_0, \ldots\}$ is R-enumerable. The program 0 HALT enumerates the empty set, as does the program

0 LET $\mathrm{R}_1 = \mathrm{R}_1 + \mathrm{a}_0$
1 GOTO 0
2 HALT.

For the sake of completeness we add the definition of register-computable functions.

2.7 Definition. Let $\mathbb{A}$ and $\mathbb{B}$ be alphabets and $F: \mathbb{A}^* \to \mathbb{B}^*$.

(a) A program P over $\mathbb{A} \cup \mathbb{B}$ *computes* F if for all $\zeta \in \mathbb{A}^*$,

$$P: \zeta \to F(\zeta).$$

(b) F is said to be *register-computable* (abbreviated: *R-computable*) if there is a program over $\mathbb{A} \cup \mathbb{B}$ which computes F.

In this terminology, program P of 2.4 computes the function $F: \{a_0, a_1\}^* \to \{a_0, a_1\}^*$ with $F(\zeta) = \zeta\zeta$. Definitions 2.5 through 2.7 can easily be extended to n-ary relations and functions. For example, in order to use a program to compute a binary function, one enters the two arguments in the first two registers.

Since any program describes a procedure it is clear that every R-decidable set is decidable, every R-enumerable set is enumerable, and every R-computable function is computable. Does the converse also hold? In other words, can every procedure in the intuitive sense be simulated by means of a program? A mathematical treatment of this problem is not possible because the concept of procedure is an intuitive one without an exact definition. Nevertheless, in spite of the simple form of the instructions allowed in register programs, it is widely accepted today that all procedures can indeed be simulated by register programs, and consequently that the intuitive concepts of decidability and enumerability coincide with their mathematically precise R-analogues. This view was first expressed by A. Church in 1936 (referring to a different but equivalent precise notion of decidability and enumerability). Therefore, the claim that every procedure can be simulated by a program and hence that the concepts of enumerability and decidability coincide with their precise counterparts is called *Church's Thesis*. We mention two arguments which support this thesis.

ARGUMENT 1: *Experience*. Hitherto it has always been possible to simulate any given procedure by a register program. In particular, programs in programming languages such as ALGOL or FORTRAN can be rewritten as register programs.

ARGUMENT 2. Since 1930 numerous mathematical concepts have been proposed as precise counterparts to the notion of procedure. Although developed from different starting points, all of these definitions have turned out to be equivalent. In the literature R-decidable sets and R-computable functions are frequently called *recursive*, and R-enumerable sets are called *recursively enumerable*.

Proofs of R-enumerability or R-decidability often require a considerable amount of programming work. To avoid getting lost in details, rather than actually writing down register programs, we shall usually content ourselves with describing procedures intuitively. The following example should help to illustrate this.

2.8 Example. The set of valid S_∞-sentences is R-enumerable.

As proof we accept the procedure described in 1.6. $\square$

In the following exercises the critical reader is invited to practice writing programs for given procedures. The more trusting reader may instead draw upon the experience of others and rely on Church's thesis.

2.9 Exercise. Suppose $W, W' \subset \mathbb{A}^*$. Show that if W and W' are R-decidable, then so are $\mathbb{A}^* - W$, $W \cap W'$, and $W \cup W'$.

2.10 Exercise. Suppose $W \subset \mathbb{A}^*$. Show:

(a) $\mathbb{A}^*$ is R-enumerable.
(b) W is R-decidable if and only if W and $\mathbb{A}^* - W$ are R-enumerable.

2.11 Exercise. Suppose $W \subset \mathbb{A}^*$. Show that (a) and (b) are equivalent.

(a) W is R-enumerable.
(b) There is a program P such that

$$P\colon \zeta \to \square, \quad \text{if } \zeta \in W.$$

$$P\colon \zeta \to \infty, \quad \text{if } \zeta \notin W.$$

2.12 Exercise. Show that a set $W \subset \mathbb{A}^*$ is R-decidable iff W is R-enumerable in lexicographical order.

§3. The Halting Problem for Register Machines

Again we fix an alphabet $\mathbb{A} = \{a_0, \ldots, a_r\}$. Our aim is to present a subset of $\mathbb{A}^*$ which is not R-decidable. The set will consist of register programs (over $\mathbb{A}$) which are suitably coded as words over $\mathbb{A}$.

For this purpose we associate with every program P (over $\mathbb{A}$) a word $\xi_P \in \mathbb{A}^*$. First we extend $\mathbb{A}$ to an alphabet $\mathbb{B}$

$$(+) \quad \mathbb{B} = \mathbb{A} \cup \{A, B, C, \ldots, X, Y, Z\} \cup \{0, 1, \ldots, 8, 9\} \cup \{=, +, -, \square, |\},$$

and we order $\mathbb{B}^*$ lexicographically according to the order of letters given in $(+)$. We represent a program P as a word over $\mathbb{B}$, e.g., the program

0 LET $\mathrm{R}_1 = \mathrm{R}_1 - \mathrm{a}_0$
1 PRINT
2 HALT

is represented by the word

$$0LETR1=R1-a_0|1PRINT|2HALT.$$

If this word is the nth word in the lexicographic ordering on $\mathbb{B}^*$, let

$$\xi_P := \underbrace{a_0 \ldots a_0}_{n\text{-times}}.$$

Set $\Pi := \{\xi_P | P \text{ is a program over } \mathbb{A}\}$.

The transition from P to ξ_P (i.e., the numbering of programs over $\mathbb{A}$ with words in $\{a_0\}^*$) is an example of a *Gödel numbering* (Gödel was the first to apply this method); and ξ_P is called the *Gödel number* of P.

Clearly, for each P we can effectively determine the corresponding $\xi_P \in \mathbb{A}^*$; conversely, given $\zeta \in \mathbb{A}^*$, we can decide whether it belongs to Π or not, and if it does we can effectively determine the program P such that $\xi_P = \zeta$. The corresponding procedures can be programmed for register machines (cf. the discussion at the end of §2). In particular we have

3.1 Lemma. *Π is R-decidable.* $\square$

The following theorem presents first examples of R-undecidable sets.

3.2 Theorem (Undecidability of the Halting Problem). (a) *The set*

$$\Pi'_{\text{halt}} = \{\xi_P | P \text{ is a program over } \mathbb{A} \text{ and } P\colon \xi_P \to \text{halt}\}$$

is not R-decidable.

(b) *The set*

$$\Pi_{\text{halt}} = \{\xi_P | P \text{ is a program over } \mathbb{A} \text{ and } P\colon \square \to \text{halt}\}$$

is not R-decidable.

Part (b) says that there is no register program which decides the set Π_{halt}. Hence by Church's Thesis there is no procedure whatsoever which decides Π_{halt}. From this we obtain the following formulation of 3.2(b):

> *There is no procedure which decides for an arbitrarily given program P whether* $P: \square \to$ *halt.*

For, if such a procedure $\mathfrak{P}$ did exist, one could use it to decide Π_{halt} as follows. First, for a given ζ, check whether $\zeta \in \Pi$ (cf. 3.1). If $\zeta \notin \Pi$ then $\zeta \notin \Pi_{\text{halt}}$. If $\zeta \in \Pi$, construct the program P for which $\xi_P = \zeta$ and then apply $\mathfrak{P}$ to P.

Proof of 3.2. (a) For a contradiction, suppose that there is a program P_0 deciding Π'_{halt}. Then for all P:

$$(1)\qquad \begin{aligned} P_0: \xi_P &\to \square, && \text{if } P: \xi_P \to \text{halt},\\ P_0: \xi_P &\to \eta \text{ for some } \eta \neq \square, && \text{if } P: \xi_P \to \infty. \end{aligned}$$

From this we easily obtain a program P_1 (see below) such that

$$(2)\qquad \begin{aligned} P_1: \xi_P &\to \infty, && \text{if } P: \xi_P \to \text{halt},\\ P_1: \xi_P &\to \text{halt}, && \text{if } P: \xi_P \to \infty. \end{aligned}$$

Then the following holds for all programs P:

$$(3)\qquad P_1: \xi_P \to \infty \quad \text{iff} \quad P: \xi_P \to \text{halt}.$$

In particular, if we set $P = P_1$, we have

$$(4)\qquad P_1: \xi_{P_1} \to \infty \quad \text{iff} \quad P_1: \xi_{P_1} \to \text{halt},$$

a contradiction.

To complete the proof of 3.2(a) we show how to construct P_1 from P_0: We change P_0 in such a way that if P_0 prints the empty word, the new program P_1 will not reach the halt instruction. This is achieved by replacing the last instruction k HALT in P_0 by

```
k    IF R₀ = □ THEN k ELSE k + 1 OR ... OR k + 1
k+1  HALT
```

and all instructions of the form L PRINT by L GOTO k.

(b) To each program P we assign in an effective way a program P^+ such that

$$(*)\qquad \begin{aligned} &P: \xi_P \to \text{halt} && \text{iff } P^+: \square \to \text{halt},\\ &\text{that is, } \xi_P \in \Pi'_{\text{halt}} && \text{iff } \xi_{P^+} \in \Pi_{\text{halt}}. \end{aligned}$$

Using $(*)$ we can prove (b) indirectly as follows:

Suppose that Π_{halt} is R-decidable, say by means of the program P_0. Then in contradiction to (a), we obtain the following decision procedure for Π'_{halt}: For an arbitrarily given $\zeta \in \mathbb{A}^*$ first check whether $\zeta \in \Pi$ (cf. 3.1). If $\zeta \notin \Pi$

then $\zeta \notin \Pi'_{\text{halt}}$. If $\zeta \in \Pi$ take the program P with Gödel number ζ, (i.e., with $\xi_P = \zeta$) and construct P^+. Using P_0, decide whether $\xi_{P^+} \in \Pi_{\text{halt}}$. On account of (*) one thus obtains an answer to the question whether $\xi_P \in \Pi'_{\text{halt}}$, i.e., whether $\zeta \in \Pi'_{\text{halt}}$.

It remains to define a program P^+ satisfying (*). If

$$\xi_P = \underbrace{a_0 \dots a_0}_{n\text{-times}}$$

let P^+ be the program which begins with the lines

$$\begin{array}{l} 0 \ \text{LET } \mathrm{R}_0 = \mathrm{R}_0 + \mathrm{a}_0 \\ \vdots \\ \mathrm{n}-1 \ \text{LET } \mathrm{R}_0 = \mathrm{R}_0 + \mathrm{a}_0 \end{array}$$

followed by the lines of P with all labels increased by n. When P^+ is started with $\square$ as input, it first builds up the word ξ_P in R_0 and then operates in the same way as the program P applied to ξ_P. Hence (*) holds. $\square$

The reader should note that the only properties of the map $P \mapsto \xi_P$ used in the proof were its injectivity and properties of effectiveness as mentioned before 3.1. Therefore the undecidability of the halting problem does not depend on our particular choice of Gödel numbering.

Of course, for particular programs P it may be easy to determine whether $P: \square \to$ halt or not. But theorem 3.2 tells us that there cannot exist a procedure which decides this question "uniformly" for each P. (Strictly speaking, 3.2 only refers to procedures which can be simulated by register programs. However, we obtain our preceding formulation if we accept Church's thesis. Henceforth we shall tacitly do this in explanatory remarks.)

The following lemma together with 3.2 shows that Π_{halt} is an example of an enumerable set which is not decidable.

3.3 Lemma. Π_{halt} *is R-enumerable.*

PROOF. We sketch an enumeration procedure: for $n = 1, 2, 3, \dots$ generate the finitely many programs whose Gödel numbers are of length $\leq n$. Start each such program with $\square$ as input, and let each one perform n steps of its computation. To compile the desired list, note each program which stops. $\square$

Applying 1.8, we obtain

3.4 Corollary. $\mathbb{A}^* - \Pi_{\text{halt}}$ *is not R-enumerable.* $\square$

The proof of 3.2(a) is based on a so-called "diagonal argument". The following exercise contains an abstract version of this method of proof.

3.5 Exercise. (a) Suppose that M is a nonempty set and $R \subset M \times M$. For $a \in M$ let

$$M_a = \{b \in M \mid Rab\}.$$

Show that the set

$$D = \{b \in M \mid \text{not } Rbb\}$$

is different from each M_a.

(b) Let $M = \mathbb{A}^*$ for some given alphabet $\mathbb{A} = \{a_0, \ldots, a_n\}$, and define $R \subset M \times M$ by

$R\xi\eta$ iff ξ is the Gödel number of a program P enumerating a set in which η occurs.

Show that

$$D = \{\eta \mid \text{not } R\eta\eta\}$$

is not enumerable. Thus the set of programs which do not print their own Gödel number is not enumerable.

(c) Let M be as in (b) and define $R \subset M \times M$ by

$R\xi\eta$ iff ξ is not the Gödel number of a program which eventually stops when started with η.

Show that all R-decidable sets ($\subset \mathbb{A}^*$) occur among the sets M_ξ and that $D = \Pi'_{\text{halt}}$.

§4. The Undecidability of First-Order Logic

The set of valid first-order S_∞-sentences is enumerable (cf. 1.6). On the other hand we have:

4.1 Theorem (Undecidability of First-Order Logic). *The set* $\{\varphi \in L_0^{S_\infty} \mid \models \varphi\}$ *of valid* S_∞*-sentences is not* R*-decidable.*

Thus there is no procedure which decides, for an arbitrary S_∞-sentence, whether it is valid or not.

Proof. We adopt the notation of §3 with $\mathbb{A} = \{|\}$. Again we identify words over $\mathbb{A}$ with natural numbers. By 3.2 we know that the set

$$\Pi_{\text{halt}} = \{\xi_P \mid P \text{ is a program over } \mathbb{A} \text{ and } P\colon \square \to \text{halt}\}$$

is not R-decidable. We shall assign to every program P, in an effective way, an S_∞-sentence φ_P such that

$$(*) \qquad \models \varphi_P \quad \text{iff } P\colon \square \to \text{halt}.$$

Then we are done: If the set $\{\varphi \in L_0^{S_\infty} \mid \models \varphi\}$ were decidable, we would have the following decision procedure for Π_{halt} (a contradiction): Given $\zeta \in \mathbb{A}^*$, first check whether ζ is of the form ξ_P. If so, take P, construct φ_P and decide whether φ_P is valid. By (*) we obtain an answer to the question whether $P: \square \to \text{halt}$, i.e., whether $\xi_P \in \Pi_{\text{halt}}$.

The following considerations are preparatory to the definition of the sentences φ_P.

Let P be a program with instructions $\alpha_0, \ldots, \alpha_k$. Denote by n the smallest number such that the registers occurring in P are among $R_0, \ldots, R_n$. An $(n+2)$-tuple $(L, m_0, \ldots, m_n)$ of natural numbers with $L \leq k$ is called a *configuration of* P. We say that $(L, m_0, \ldots, m_n)$ is the *configuration of* P *after* s *steps* if P, started with $\square$, runs for at least s steps, and if after s steps instruction L is to be executed next, while the numbers $m_0, \ldots, m_n$ are in $R_0, \ldots, R_n$, respectively. In particular, $(0, 0, \ldots, 0)$ is the configuration of P after 0 steps (the "initial configuration"). Since only α_k is a halt-instruction we have

(1) $P: \square \to \text{halt}$ iff for suitable $s, m_0, \ldots, m_n$, $(k, m_0, \ldots, m_n)$ is the configuration of P after s steps.

If $P: \square \to \text{halt}$, we let s_P be the number of steps carried out by P until it arrives at the halt-instruction. Finally we choose symbols R ($(n+3)$-ary), $<$ (binary), f (unary), and c from S_∞ (e.g., R_0^{n+3}, R_0^2, f_0^1, and c_0), and set $S := \{R, <, f, c\}$.

With the program P we associate an S-structure $\mathfrak{A}_P$ within which we shall describe how P operates. We distinguish two cases:

Case 1. $P: \square \to \infty$. We set $A_P := \mathbb{N}$ and interpret $<$ by the usual ordering on $\mathbb{N}$, c by 0, f by the successor function, and R by the relation $\{(s, L, m_0, \ldots, m_n) \mid (L, m_0, \ldots, m_n)$ is the configuration of P after s steps$\}$, respectively.

Case 2. $P: \square \to \text{halt}$. We set $e := \max\{k, s_P\}$ and $A_P := \{0, \ldots, e\}$, and interpret $<$ by the usual ordering on A_P and c by 0, respectively; furthermore we define f^{A_P} by $f^{A_P}(m) = m + 1$ for $m < e$ and $f^{A_P}(e) = e$, and set $R^{A_P} := \{(s, L, m_0, \ldots, m_n) \mid (L, m_0, \ldots, m_n)$ is the configuration of P after s steps$\}$. (Note that R^{A_P} is indeed a relation on A_P since at each step P increases the contents of each register by at most 1, and hence we have $m_0, \ldots, m_n \leq s_P \leq e$ for all $(s, L, m_0, \ldots, m_n) \in R^{A_P}$.)

Now we provide an S-sentence ψ_P which, in a suitable way, describes the operations of P on $\square$. We abbreviate $c, fc, ffc, \ldots$ by $\bar{0}, \bar{1}, \bar{2}, \ldots$, respectively. In reading ψ_P the reader should convince himself that the following holds:

(2) (a) $\mathfrak{A}_P \models \psi_P$.

(b) If $\mathfrak{A}$ is an S-structure with $\mathfrak{A} \models \psi_P$ and $(L, m_0, \ldots, m_n)$ is the configuration of P after s steps, then the elements $\bar{0}^A, \bar{1}^A, \ldots, \bar{s}^A$ are pairwise distinct and $\mathfrak{A} \models R\bar{s}\bar{L}\bar{m}_0 \ldots \bar{m}_n$.

We set

$$\psi_P := \psi_0 \wedge R\bar{0}\dots\bar{0} \wedge \psi_{\alpha_0} \wedge \dots \wedge \psi_{\alpha_{k-1}},$$

where $\psi_0, \psi_{\alpha_0}, \dots, \psi_{\alpha_{k-1}}$ will now be defined. ψ_0 says that $<$ is an ordering whose first element is c, that $x \leq fx$ holds for every x, and that fx is the immediate successor of x in case x is not the last element,

$$\begin{aligned}\psi_0 := \text{“} < \text{is an ordering”} &\wedge \forall x(c < x \vee c \equiv x)\\ &\wedge \forall x(x < fx \vee x \equiv fx) \wedge \forall x(\exists y\, x < y \rightarrow (x < fx\\ &\qquad \wedge \forall z(x < z \rightarrow (fx < z \vee fx \equiv z)))).\end{aligned}$$

For $\alpha = \alpha_0, \dots, \alpha_{k-1}$, the sentence ψ_α describes the operation corresponding to instruction α. It is defined as follows.

For $\alpha = \mathrm{L}$ LET $\mathrm{R_i} = \mathrm{R_i} + |$:

$$\begin{aligned}\psi_\alpha := \forall x\, \forall y_0 \dots \forall y_n(Rx\bar{L}y_0 \dots y_n \rightarrow (x < fx\\ \wedge Rfx\overline{L+1}y_0 \dots y_{i-1}fy_iy_{i+1} \dots y_n).\end{aligned}$$

For $\alpha = \mathrm{L}$ LET $\mathrm{R_i} = \mathrm{R_i} - |$:

$$\begin{aligned}\psi_\alpha := \forall x\, \forall y_0 \dots \forall y_n(Rx\bar{L}y_0 \dots y_n \rightarrow (x < fx\\ \wedge ((y_i \equiv \bar{0} \wedge Rfx\overline{L+1}y_0 \dots y_n) \vee (\neg y_i \equiv \bar{0}\\ \wedge \exists u(fu = y_i \wedge Rfx\overline{L+1}y_0 \dots y_{i-1}uy_{i+1} \dots y_n))))).\end{aligned}$$

For $\alpha = \mathrm{L}$ IF $\mathrm{R_i} = \square$ THEN L' ELSE $\mathrm{L_0}$:

$$\begin{aligned}\psi_\alpha := \forall x\, \forall y_0 \dots \forall y_n(Rx\bar{L}y_0 \dots y_n \rightarrow (x < fx\\ \wedge ((y_i \equiv \bar{0} \wedge Rfx\bar{L}'y_0 \dots y_n) \vee (\neg y_i \equiv \bar{0} \wedge Rfx\bar{L}_0y_0 \dots y_n)))).\end{aligned}$$

For $\alpha = \mathrm{L}$ PRINT:

$$\psi_\alpha := \forall x\, \forall y_0 \dots \forall y_n(Rx\bar{L}y_0 \dots y_n \rightarrow (x < fx \wedge Rfx\overline{L+1}y_0 \dots y_n)).$$

Now we set

(3) $\varphi_P := \psi_P \rightarrow \exists x\, \exists y_0 \dots \exists y_n\, Rx\bar{k}y_0 \dots y_n.$

Then φ_P is an S-sentence which satisfies (*), i.e.,

$$\models \varphi_P \quad \text{iff } P\colon \square \rightarrow \text{halt}.$$

Indeed, suppose first that φ_P is valid. Then in particular $\mathfrak{A}_P \models \varphi_P$. Since $\mathfrak{A}_P \models \psi_P$ (cf. (2)(a)), we have $\mathfrak{A}_P \models \exists x\, \exists y_0 \dots \exists y_n\, Rx\bar{k}y_0 \dots y_n$ (cf. (3)), i.e., for suitable $s, m_0, \dots, m_n$, $(k, m_0, \dots, m_n)$ is the configuration of P after s steps. Now, (1) yields $P\colon \square \rightarrow$ halt.

Conversely, if $P\colon \square \rightarrow$ halt, then for suitable $s, m_0, \dots, m_n$, the tuple $(k, m_0, \dots, m_n)$ is the configuration of P after s steps. Hence φ_P is valid, because if $\mathfrak{A}$ is an S-structure such that $\mathfrak{A} \models \psi_P$, then $\mathfrak{A} \models R\bar{s}\bar{k}\bar{m}_0 \dots \bar{m}_n$ by (2)(b) and hence $\mathfrak{A} \models \varphi_P$. $\square$

The undecidability of first-order logic was first proved by A. Church [5] in 1936. In traditional logic (Llull, Leibniz) the problem of finding a decision procedure for "logically true propositions" had already been considered. 4.1 shows that such a search was bound to fail.

4.2 Exercise. Prove (2)(b) by induction over s.

4.3 Exercise. Show that the set of satisfiable S_∞-sentences is not R-enumerable.

§5. Trahtenbrot's Theorem and the Incompleteness of Second-Order Logic

The object of this section is to prove that the set of valid second-order S_∞-sentences is not enumerable, and to briefly discuss the methodological consequences. A useful tool in this context will be Trahtenbrot's theorem, which says that the set of sentences valid in all finite structures is not enumerable.

5.1 Definition. (a) An S-sentence φ is said to be *fin-satisfiable* if there is a finite S-structure which satisfies φ.
(b) An S-sentence φ is said to be *fin-valid* if every finite S-structure satisfies φ.

To be specific we consider the symbol set S_∞ and put

$$\Phi_{\mathrm{fs}} := \{\varphi \in L_0^{S_\infty} \mid \varphi \text{ fin-satisfiable}\},$$

$$\Phi_{\mathrm{fv}} := \{\varphi \in L_0^{S_\infty} \mid \varphi \text{ fin-valid}\}.$$

As an example, we note that over a finite domain any injective function is also surjective; therefore the sentence $\varphi := \forall x\, \forall y (fx \equiv fy \rightarrow x \equiv y) \rightarrow \forall x\, \exists y\, x \equiv fy$ is fin-valid; however, φ is not valid. The sentence $\neg\varphi$ is satisfiable but not fin-satisfiable.

5.2 Lemma. *Φ_{fs} is R-enumerable.*

Proof. First we describe a procedure which decides, for every S_∞-sentence φ and every $n \geq 1$, whether or not φ is satisfiable over a domain with n elements. Suppose φ and n are given. Since for every structure with n elements there is an isomorphic structure with domain $\{1, \ldots, n\}$, we only need to check (by the isomorphism lemma) whether φ is satisfiable over $\{1, \ldots, n\}$. Let S be the (finite!) set of symbols occurring in φ and $\mathfrak{A}_0, \ldots, \mathfrak{A}_k$ be the finitely many S-structures with domain $\{1, \ldots, n\}$ (cf. III.1.5). We can describe the $\mathfrak{A}_i$ explicitly by means of finite tables for the relations, functions and constants. φ is satisfiable over $\{1, \ldots, n\}$ if and only if $\mathfrak{A}_i \models \varphi$ for some

$i < k$. Thus we only need to test whether $\mathfrak{A}_i \models \varphi$ for $i = 0, \ldots, k$. These tests can be reduced to questions which can be answered from the respective tables as follows: Given $i < k$, if $\varphi = \neg\psi$ then the problem "$\mathfrak{A}_i \models \varphi$?" can be reduced to the question of whether $\mathfrak{A}_i \models \psi$. If $\varphi = (\psi \vee \chi)$ then similarly the problem can be reduced to the questions of whether $\mathfrak{A}_i \models \psi$ and whether $\mathfrak{A}_i \models \chi$. If $\varphi = \exists x\psi$ we reduce to the questions "$\mathfrak{A}_i \models \psi[1]$?", ..., "$\mathfrak{A}_i \models \psi[n]$?". Continuing in this way we eventually arrive at questions of the form "$\mathfrak{A}_i \models \psi[n_0, \ldots, n_{m-1}]$?" for atomic formulas $\psi(v_0, \ldots, v_{m-1})$ and $n_0, \ldots, n_{m-1} \leq n$. Clearly these can be answered effectively by inspecting the tables for $\mathfrak{A}_i$.

Now Φ_{fs} can be enumerated as follows: For $m = 1, 2, \ldots$ generate the (finitely many) words over $\mathbb{A}_0$ which are S_∞-sentences and are of length $\leq m$, and use the procedure just described to decide, for $n = 1, \ldots, m$, whether they are satisfiable over a domain with n elements. List the sentences for which this is the case. □

5.3 Theorem. *Φ_{fs} is not R-decidable.*

Proof. For a program P over $\mathbb{A} = \{|\}$, let $\mathfrak{A}_P$ and ψ_P be defined as in the proof of 4.1. We show

$$(*) \qquad P\colon \square \to \text{halt} \quad \text{iff } \psi_P \in \Phi_{\text{fs}}.$$

This proves the theorem; for otherwise, using (*), one could obtain from a decision procedure for Φ_{fs} a procedure to decide whether $P\colon \square \to \text{halt}$ (cf. the corresponding argument in the proof of 4.1).

Proof of (*). If $P\colon \square \to \text{halt}$, then $\mathfrak{A}_P$ is finite and is a model of ψ_P. Hence $\psi_P \in \Phi_{\text{fs}}$. Conversely, if $P\colon \square \to \infty$, then by (2)(b) in the proof of 4.1, the elements $\bar{0}^{\mathfrak{A}}, \bar{1}^{\mathfrak{A}}, \ldots$ are pairwise distinct in every model $\mathfrak{A}$ of ψ_P. Thus every model of ψ_P is infinite, and hence $\psi_P \notin \Phi_{\text{fs}}$. □

From 5.2 and 5.3 we now obtain

5.4 Trahtenbrot's Theorem. *The set Φ_{fv} of S_∞-sentences valid in all finite structures is not R-enumerable.*

Proof. Clearly,

$$(*) \qquad \varphi \in L_0^{S_\infty} - \Phi_{\text{fs}} \quad \text{iff } \neg\varphi \in \Phi_{\text{fv}}$$

holds for $\varphi \in L_0^{S_\infty}$. For a contradiction assume that Φ_{fv} is enumerable. Then, using (*), one can also enumerate $L_0^{S_\infty} - \Phi_{\text{fs}}$: one simply starts an enumeration procedure for Φ_{fv}, and whenever it lists a sentence $\neg\varphi$, one writes down φ. This would lead to a decision procedure for Φ_{fs} (in contradiction to 5.3) as follows: For a string ζ over $\mathbb{A}_0$, decide first whether ζ is an S_∞-sentence. If so, start enumeration procedures for Φ_{fs} (cf. 5.2) and for $L_0^{S_\infty} - \Phi_{\text{fs}}$, and let both procedures continue until one of them yields ζ as output. Thus one obtains a decision whether $\zeta \in \Phi_{\text{fs}}$. □

5.5 Theorem (Incompleteness of Second-Order Logic). *The set of valid second-order S_∞-sentences is not R-enumerable.*

PROOF. Let φ_{fin} be a second-order S_∞-sentence with the property that for all $\mathfrak{A}$,

$$\mathfrak{A} \models \varphi_{\text{fin}} \quad \text{iff } \mathfrak{A} \text{ is finite}$$

(cf. IX.1). Then for all first-order S_∞-sentences φ,

$$(*) \qquad \varphi \in \Phi_{\text{fv}} \quad \text{iff } \models \varphi_{\text{fin}} \to \varphi.$$

Now, if the set of valid second-order S_∞-sentences is R-enumerable, then one can start an enumeration procedure for this set, and each time it yields a sentence of the form $\varphi_{\text{fin}} \to \varphi$, where $\varphi \in L_0^{S_\infty}$, one adds φ to the list. By (*) we obtain in this way an enumeration of Φ_{fv}, in contradiction to Trahtenbrot's theorem. □

Theorem 5.5 is due to Gödel. It is a stronger version of a result obtained in IX.1. There we concluded from the failure of the compactness theorem for second-order logic $\mathscr{L}_{\text{II}}$, that there cannot be any correct and complete proof calculus for $\mathscr{L}_{\text{II}}$. In other words, there is no calculus whose derivability relation $\vdash$ satisfies

(+) For all $\mathscr{L}_{\text{II}}$-sentences φ and all sets Φ of $\mathscr{L}_{\text{II}}$-sentences, $\Phi \models \varphi$ iff $\Phi \vdash \varphi$.

However, (+) leaves open the question of whether there is a calculus which satisfies (+) for $\Phi = \emptyset$, that is, whether there is a correct calculus in which all valid second-order sentences are derivable. Now 5.5 shows that in this sense second-order logic is also incomplete: If such a calculus existed, one could apply its rules systematically to generate all possible derivations and hence all valid second-order sentences (cf. the proof of 1.6).

At this point we see how useful it has been to introduce the notion of enumerability: By employing this notion we were relieved of the task of giving precise definitions for the notions of derivation rule and calculus, but were nevertheless able to conclude that there is no adequate proof calculus for the valid second-order sentences.

The argument for 5.5 above is based on the fact that the finite sets are characterizable in second-order logic. Thus it can also be applied to weak second-order logic (cf. IX.1.7).

For the sake of simplicity, we have, in the last section, referred to the symbol set S_∞ although we have actually needed only a few symbols from S_∞. It should be clear that the results are also valid for other symbol sets S which are effectively given as is S_∞ and contain the symbols mentioned above. One can even show that it is sufficient for S to contain only one binary relation symbol. Moreover, the incompleteness of second-order logic already holds for $S = \emptyset$ (cf. 5.6). On the other hand, the set of valid first-order S-sentences is decidable provided S contains only unary relation symbols (cf. XI.3.9(b)).

5.6 Exercise. The set of valid second-order $\emptyset$-sentences is not R-enumerable.

§6. Theories and Decidability

In this section we investigate several theories, especially with regard to enumerability and decidability. Among the results obtained is the undecidability of arithmetic. We shall always assume that the symbol sets considered are effectively given.

A. First-Order Theories

6.1 Definition. $T \subset L_0^S$ is said to be a *theory* if T is satisfiable and if it is closed under consequence (i.e., every S-sentence which follows from T already belongs to T).

For every S-structure $\mathfrak{A}$ the set

$$\mathrm{Th}(\mathfrak{A}) := \{\varphi \in L_0^S \mid \mathfrak{A} \models \varphi\}$$

is a theory, the *theory of* $\mathfrak{A}$ (cf. VI.4.1). $\mathrm{Th}(\mathfrak{N})$ is called (elementary) *arithmetic*.

For $\Phi \subset L_0^S$ let $\Phi^{\models} := \{\varphi \in L_0^S \mid \Phi \models \varphi\}$. If T is a theory, then $T = T^{\models}$, and if Φ is a satisfiable set of S-sentences, then $\Phi^{\models}$ is a theory. We give a few examples.

(1) $\emptyset^{\models} = \{\varphi \in L_0^S \mid \models \varphi\}$.
(2) For $S = S_{\mathrm{gr}}$: (first-order) group theory $\mathrm{Th}_{\mathrm{gr}} = \Phi_{\mathrm{gr}}^{\models}$.
(3) For $S = \{\in\}$: ZFC set theory $\mathrm{Th}_{\mathrm{ZFC}} := \mathrm{ZFC}^{\models}$.
(4) For $S = S_{\mathrm{ar}}$: the so-called (first-order) *Peano arithmetic* $\mathrm{Th}_{\mathrm{PA}} := \Phi_{\mathrm{PA}}^{\models}$. The axiom system Φ_{PA} consists of the Peano axioms given in III.7.5, where the usual induction axiom (a second-order sentence) is replaced by the first-order "induction axioms" (*) below:

$$\forall x \neg x + 1 \equiv 0, \quad \forall x \forall y(x + 1 \equiv y + 1 \rightarrow x \equiv y),$$

$$\forall x x + 0 \equiv x, \quad \forall x \forall y\, x + (y + 1) \equiv (x + y) + 1,$$

$$\forall x x \cdot 0 \equiv 0, \quad \forall x \forall y x \cdot (y + 1) \equiv x \cdot y + x,$$

(*) and for all $x_0, \ldots, x_{n-1}$, y and all $\varphi \in L^{S_{\mathrm{ar}}}$ such that $\mathrm{free}(\varphi) \subset \{x_0, \ldots, x_{n-1}, y\}$ the sentence

$$\forall x_0 \ldots \forall x_{n-1}\left(\left(\varphi\frac{0}{y} \wedge \forall y\left(\varphi \rightarrow \varphi\frac{y+1}{y}\right)\right) \rightarrow \forall y \varphi\right).$$

$\mathfrak{N}$ is a model of Φ_{PA}. The schema (*) is a natural substitute for the induction axiom, because it expresses the induction axiom for properties which are definable in first-order logic. Many theorems of elementary arithmetic (i.e., sentences in $\mathrm{Th}(\mathfrak{N})$) can be derived from Φ_{PA}. Nevertheless it turns out that not *all* sentences of $\mathrm{Th}(\mathfrak{N})$ are derivable from Φ_{PA}: in 6.10 we shall show that $\Phi_{PA}^{\models} \subsetneqq \mathrm{Th}(\mathfrak{N})$.

6.2 Definition. (a) A theory T is said to be *R-axiomatizable* if there is an R-decidable set Φ of sentences such that $T = \Phi^{\models}$.
(b) A theory T is said to be *finitely axiomatizable* if there is a finite set Φ of sentences such that $T = \Phi^{\models}$.

Every finitely axiomatizable theory can be axiomatized by means of a single sentence. (Take the conjunction of the axioms.) Every finitely axiomatizable theory is also R-axiomatizable. The theories Th_{PA} and Th_{ZFC} are R-axiomatizable, but it can be shown that they are not finitely axiomatizable.

6.3 Theorem. *An R-axiomatizable theory is R-enumerable.*

Proof. Let T be a theory and let Φ be an R-decidable set of S-sentences such that $T = \Phi^{\models}$. The sentences of T may be listed as follows: Generate systematically all derivable sequents and check in each case whether the members of the antecedent belong to Φ. If so, list the succedent provided it is a sentence. □

An R-axiomatizable theory T need not necessarily be R-decidable. Examples are $T = \emptyset^{\models}$ (for $S = S_\infty$; cf. 4.1) and $T = \mathrm{Th}_{gr}$ (cf. [28]). The situation is different, however, if T is complete in the following sense.

6.4 Definition. A theory $T \subset L_0^S$ is *complete* if for every S-sentence φ we have $\varphi \in T$ or $\neg\varphi \in T$.

$\mathrm{Th}(\mathfrak{A})$ is complete for every structure $\mathfrak{A}$.

6.5 Theorem. (a) *Every R-axiomatizable and complete theory is R-decidable.*
(b) *Every R-enumerable and complete theory is R-decidable.*

Proof. By 6.3 it is sufficient to prove (b). Let T be an R-enumerable complete theory. In order to decide whether a given sentence φ belongs to T, we use a procedure to enumerate T, continuing until either φ or $\neg\varphi$ has been listed. Since T is complete, one of these two sentences will eventually be listed. If φ is listed, φ belongs to T; if $\neg\varphi$ is listed, φ does not belong to T. □

From 6.5 we obtain the decidability of an axiomatizable theory once we have proved its completeness. A method for proving completeness will be introduced in the next chapter. In certain cases one can use the assertion in 6.7 for this purpose.

6.6 Exercise. Let $T = \Phi^{\models}$ be a theory, where Φ is R-enumerable. Show that T is R-axiomatizable. (*Hint*: Starting with an enumeration $\varphi_0, \varphi_1, \ldots$ of Φ, consider the set $\{\varphi_0, \varphi_0 \wedge \varphi_1, \ldots\}$.)

6.7 Exercise. (a) For at most countable S, let $T \subset L_0^S$ be a theory having only infinite models. Further, suppose there is an infinite cardinal κ such that any two models of T of cardinality κ are isomorphic. Show that T is complete.

(b) Set up a decidable system of axioms for the theory of algebraically closed fields of fixed characteristic and use (a) to show its completeness (and hence by 6.5 its decidability).

B. The Undecidability of Arithmetic

In this section we prove the undecidability of arithmetic, i.e., we show that there is no procedure which decides for every S_{ar}-sentence whether it holds in $\mathfrak{N}$. We shall use the same method of proof as in showing the undecidability of first-order logic: we effectively assign to every register program P an S_{ar}-sentence φ_P such that

$$\mathfrak{N} \models \varphi_P \quad \text{iff } P\colon \square \to \text{halt}.$$

The undecidability of $\mathrm{Th}(\mathfrak{N})$ then follows immediately from the undecidability of Π_{halt}.

In defining φ_P we shall make use of a formula ψ_P which, in $\mathfrak{N}$, describes how the program P operates. The following lemma provides us with such a formula.

Assume the program P consists of the instructions $\alpha_0, \ldots, \alpha_k$, and let n be the smallest number such that all registers mentioned in P are among $R_0, \ldots, R_n$. Recall (cf. §4) that a configuration of P is an $(n + 2)$-tuple $(L, m_0, \ldots, m_n)$ of natural numbers such that $L \leq k$. $(L, m_0, \ldots, m_n)$ stands for a situation where α_L is the next instruction to be executed and the contents of the registers are $m_0, \ldots, m_n$.

6.8 Lemma. *With any given program P one can effectively associate a formula $\psi_P(v_0, \ldots, v_{2n+2})$ such that for all $k_0, \ldots, k_n, L, m_0, \ldots, m_n \in \mathbb{N}$ the following holds*:

> $\mathfrak{N} \models \psi_P[k_0, \ldots, k_n, L, m_0, \ldots, m_n]$ *iff P, beginning with the configuration $(0, k_0, \ldots, k_n)$, after finitely many steps reaches the configuration $(L, m_0, \ldots, m_n)$.*

Using ψ_P, we can write down the desired formula φ_P as

$$\varphi_P := \exists v_{n+2} \ldots \exists v_{2n+2}\, \psi_P(\underline{0}, \ldots, \underline{0}, \underline{k}, v_{n+2}, \ldots, v_{2n+2}).^1$$

[1] In case $\varphi \in L_2^{S_{\mathrm{ar}}}$, for example, we write $\varphi(\underline{n}, v_1)$ for $\varphi \frac{\underline{n}}{v_0}$ and $\varphi(\underline{n}, \underline{m})$ for $\varphi \frac{\underline{n}\,\underline{m}}{v_0\, v_1}$. Here, as before, $\underline{n}$ denotes the corresponding term $1 + \cdots + 1$.

Then we have (note that α_k is the halt-instruction of P):

$\mathfrak{N} \models \varphi_P$ iff P, beginning with the configuration $(0, \ldots, 0)$, after finitely many steps reaches the configuration $(k, m_0, \ldots, m_n)$ for some $m_0, \ldots, m_n$
iff $P: \square \to$ halt.

Thus we have

6.9 Theorem (Undecidability of Arithmetic). *Arithmetic, i.e., the S_{ar}-theory* Th($\mathfrak{N}$), *is not R-decidable.* □

Since Th($\mathfrak{N}$) is complete, using 6.5, we obtain

6.10 Corollary. *Arithmetic is neither R-axiomatizable nor R-enumerable.* □

According to 6.9 and 6.10 arithmetic is not amenable to a purely "mechanical" treatment in the following sense: There is no procedure for deciding whether any given arithmetical sentence is true, nor is there even a procedure which lists all true arithmetical sentences. In other words, every procedure which lists only true arithmetical sentences must necessarily omit some true arithmetical sentences. Thus mathematicians will never possess a method for systematically proving all true arithmetical sentences.

Proof of Lemma 6.8. Let P be given as above. We must find an S_{ar}-formula $\psi_P(x_0, \ldots, x_n, z, y_0, \ldots, y_n)$ $(= \psi_P(\bar{x}, z, \bar{y}))$ which says (in $\mathfrak{N}$) that P, beginning with the configuration $(0, \bar{x})$, proceeds through a series of configurations, ending finally with the configuration $(z, \bar{y})$. That is, $\psi_P(\bar{x}, z, \bar{y})$ should be a formalization of the following statement:

"There is an $s \in \mathbb{N}$ and a sequence $C_0, \ldots, C_s$ of configurations such that

(1) $\quad C_0 = (0, x_0, \ldots, x_n), \qquad C_s = (z, y_0, \ldots, y_n),$

and for all $i < s$, $C_i \underset{P}{\to} C_{i+1}$."

"$C_i \underset{P}{\to} C_{i+1}$" means that P passes from configuration C_i to C_{i+1} when executing the instruction addressed in C_i. We form a single sequence from $C_0, \ldots, C_s$ and thus obtain the following formulation of (1):

"There is an $s \in \mathbb{N}$ and a sequence

$$(\underbrace{a_0, \ldots, a_{n+1}}_{C_0}, \underbrace{a_{n+2}, \ldots, a_{(n+2)+(n+1)}}_{C_1}, \ldots, \underbrace{a_{s\cdot(n+2)}, \ldots, a_{s\cdot(n+2)+(n+1)}}_{C_s})$$

(2) such that

$$a_0 = 0, a_1 = x_0, \ldots, a_{n+1} = x_n,$$

$$a_{s(n+2)} = z, a_{s\cdot(n+2)+1} = y_0, \ldots, a_{s\cdot(n+2)+(n+1)} = y_n,$$

and for all $i < s$,

$$(a_{i\cdot(n+2)}, \ldots, a_{i\cdot(n+2)+(n+1)}) \underset{P}{\rightarrow} (a_{(i+1)\cdot(n+2)}, \ldots, a_{(i+1)\cdot(n+2)+(n+1)})".$$

The principal difficulty in formalizing (2) as a first-order S_{ar}-sentence arises with the quantifier "there exists a sequence...". We overcome this problem by using natural numbers as codes for finite sequences. Often one codes a sequence $(a_0, \ldots, a_r)$ by the number $p_0^{a_0+1} \cdot \ldots \cdot p_r^{a_r+1}$, where p_i denotes the ith prime. However, when using this code, we would be forced to give an $L^{S_{\text{ar}}}$-definition of exponentiation. Since such a definition is rather involved, we provide another coding where a sequence $(a_0, \ldots, a_r)$ is coded by two suitably chosen numbers t and p.

6.11 β-function Lemma.[2] *There is a function β: $\mathbb{N}^3 \rightarrow \mathbb{N}$ with the following properties:*

(a) *For every sequence $(a_0, \ldots, a_r)$ over $\mathbb{N}$, there exist $t, p \in \mathbb{N}$ such that for $i \leq r$*

$$\beta(t, p, i) = a_i.$$

(b) *There exists an S_{ar}-formula $\chi(v_0, v_1, v_2, v_3)$ which defines β in $\mathfrak{N}$ in the sense that for $t, p, i, a \in \mathbb{N}$,*

$$\mathfrak{N} \models \chi[t, p, i, a] \quad \textit{iff} \quad \beta(t, p, i) = a.$$

Proof. (a) Given $(a_0, \ldots, a_r)$, we choose a prime p which is larger than $a_0, \ldots, a_r, r + 1$ and set

$$(*) \qquad t := 1 \cdot p^0 + a_0 p^1 + 2p^2 + a_1 p^3 + \cdots + (r + 1)p^{2r} + a_r p^{2r+1}.$$

By choice of p the right-hand side is the p-adic representation of t.

First we show that for all $i \leq r$

$$(**) \qquad a = a_i \text{ iff there are } b_0, b_1, b_2 \text{ such that } \begin{array}{ll} \text{(i)} & t = b_0 + b_1((i + 1) + ap + b_2 p^2), \\ \text{(ii)} & a < p, \\ \text{(iii)} & b_0 < b_1, \\ \text{(iv)} & b_1 = p^{2l} \text{ for a suitable } l. \end{array}$$

The implication from left to right follows immediately from $(*)$. Conversely, suppose (i)–(iv) hold for b_0, b_1, b_2 and let $b_1 = p^{2l}$. From (i) we obtain

$$t = b_0 + (i + 1) \cdot p^{2l} + ap^{2l+1} + b_2 p^{2l+2}.$$

[2] This nomenclature stems from Gödel's use of β for a function with the properties (a) and (b) of the lemma.

Since $b_0 < p^{2l}$, $a < p$, and the p-adic representation of t is unique, a comparison with (*) yields $l = i$ and $a = a_i$. We set $\beta(t, p, i) = a_i$, i.e., the uniquely determined number a for which the right-hand side of (**) holds. We extend this definition to arbitrary natural numbers r, q, j by specifying

$$\beta(r, q, j) = \begin{cases} \text{the smallest } a \text{ such that there exist } b_0, b_1, b_2 \text{ with} \\ \quad \text{(i) } r = b_0 + b_1((j + 1) + aq + b_2 q^2) \\ \quad \text{(ii) } a < q \\ \quad \text{(iii) } b_0 < b_1 \\ \quad \text{(iv) } b_1 = q^{2l} \quad \text{for suitable } l, \\ \quad \text{if such an } a \text{ exists and } q \text{ is prime} \\ 0, \quad \text{otherwise.} \end{cases}$$

Then β has the properties required in (a).

(b) The definition just given leads immediately to an S_{ar}-formula $\chi(v_0, v_1, v_2, v_3)$ defining β; one need only note that (iv) is equivalent to the condition that b_1 be a square and that for all $d \neq 1$ with $d \mid b_1$ we have $q \mid d$. □

We now return to the proof of 6.8, that is, to the problem of giving an S_{ar}-formula which says that the program P passes in finitely many steps from the configuration $(0, \bar{x})$ to the configuration $(z, \bar{y})$. As we have seen, this statement about P is equivalent to statement (2). We can formalize (2) with the aid of the formula χ from the β-function lemma (where we now use $s, t, \ldots$ to denote variables):

$$\begin{aligned}
&\psi_P(x_0, \ldots, x_n, z, y_0, \ldots, y_n) \\
&\quad := \exists s\, \exists p\, \exists t(\chi(t, p, \underline{0}, \underline{0}) \wedge \chi(t, p, \underline{1}, x_0) \wedge \cdots \wedge \chi(t, p, \underline{n+1}, x_n) \\
&\quad \wedge \chi(t, p, s \cdot (\underline{n+2}), z) \wedge \chi(t, p, s \cdot (\underline{n+2}) + 1, y_0) \wedge \cdots \\
&\qquad\qquad \wedge \chi(t, p, s \cdot (\underline{n+2}) + (\underline{n+1}), y_n) \\
&\quad \wedge \forall i\, \forall u u_0 \ldots u_n\, \forall u' u'_0 \ldots u'_n \\
&\qquad [i < s \wedge \chi(t, p, i \cdot (\underline{n+2}), u) \wedge \chi(t, p, i \cdot (\underline{n+2}) + 1, u_0) \wedge \cdots \\
&\qquad\qquad \wedge \chi(t, p, i \cdot (\underline{n+2}) + (\underline{n+1}), u_n) \\
&\quad \wedge \chi(t, p, (i+1) \cdot (\underline{n+2}), u') \wedge \chi(t, p, (i+1) \cdot (\underline{n+2}) + 1, u'_0) \wedge \cdots \\
&\qquad\qquad \wedge \chi(t, p, (i+1) \cdot (\underline{n+2}) + (\underline{n+1}), u'_n) \\
&\quad \to \text{“}(u, u_0, \ldots, u_n) \underset{P}{\to} (u', u'_0, \ldots, u'_n)\text{”}]).
\end{aligned}$$

Here

$$\text{“}(u, u_0, \ldots, u_n) \underset{P}{\to} (u', u'_0, \ldots, u'_n)\text{”}$$

stands for a formula which describes the direct transition from configuration $(u, u_0, \ldots, u_n)$ to configuration $(u', u'_0, \ldots, u'_n)$; such a formula can be obtained as a conjunction $\psi_0 \wedge \cdots \wedge \psi_{k-1}$, where ψ_j describes transitions induced by instruction α_j of P. For example, if α_j is of the form

$$\text{j LET } R_1 = R_1 + 1$$

then we take

$$\psi_j := u \equiv \underline{j} \to (u' \equiv u + 1 \wedge u_0' \equiv u_0 \wedge u_1' \equiv u_1 + 1 \wedge u_2' \equiv u_2 \wedge \cdots \wedge u_n' \equiv u_n).$$

Instructions of other type can be treated similarly. Thus a formula ψ_p with the desired properties is obtained, and the proof of 6.8 is completed. □

Finally, we note another consequence of the fact that computations of register machines can be described in $\mathfrak{N}$.

6.12 Theorem. *Let* $r \geq 1$.

(a) *Given an r-ary R-decidable relation* $\mathfrak{Q}$ *over* $\mathbb{N}$, *there is an* S_{ar}*-formula* $\varphi(v_0, \ldots, v_{r-1})$ *such that for all* $k_0, \ldots, k_{r-1} \in \mathbb{N}$,

$$\mathfrak{Q}k_0 \ldots k_{r-1} \quad \text{iff } \mathfrak{N} \models \varphi(\underline{k}_0, \ldots, \underline{k}_{r-1}).$$

(b) *Given an R-computable function* $f: \mathbb{N}^r \to \mathbb{N}$, *there is an* S_{ar}*-formula* $\varphi(v_0, \ldots, v_{r-1}, v_r)$ *such that for all* $k_0, \ldots, k_{r-1}, k_r \in \mathbb{N}$,

$$f(k_0, \ldots, k_{r-1}) = k_r \quad \text{iff } \mathfrak{N} \models \varphi(\underline{k}_0, \ldots, \underline{k}_{r-1}, \underline{k}_r),$$

and in particular,

$$\mathfrak{N} \models \exists^{=1} v_r \varphi(\underline{k}_0, \ldots, \underline{k}_{r-1}, v_r).$$

Proof. (a) Suppose $r \geq 1$ and let $\mathfrak{Q}$ be an r-ary R-decidable relation over $\mathbb{N}$. Let P be a register program which decides $\mathfrak{Q}$ and $\alpha_{L_0}, \ldots, \alpha_{L_l}$ be the print-instructions of P. Suppose that R_n is the largest register mentioned in P, and without loss of generality, that $n \geq r - 1$. Then, using ψ_P from 6.8, we have for arbitrary $k_0, \ldots, k_{r-1} \in \mathbb{N}$:

$\mathfrak{Q}k_0 \ldots k_{r-1}$ iff P, beginning with configuration

$$(0, k_0, \ldots, k_{r-1}, \underbrace{0, \ldots, 0}_{n-r+1}),$$

after finitely many steps reaches a configuration of the form $(L_i, 0, m_1, \ldots, m_n)$ with $0 \leq i \leq l$

iff $\mathfrak{N} \models \exists v_{n+3} \ldots v_{2n+2}$

$$(\psi_P(\underline{k}_0, \ldots, \underline{k}_{r-1}, \underline{0}, \ldots, \underline{0}, \underline{L}_0, \underline{0}, v_{n+3}, \ldots, v_{2n+2}) \vee \cdots \vee \psi_P(\underline{k}_0, \ldots, \underline{k}_{r-1}, \underline{0}, \ldots, \underline{0}, \underline{L}_l, \underline{0}, v_{n+3}, \ldots, v_{2n+2})).$$

Thus for $\varphi(v_0, \ldots, v_{r-1})$ one can take the formula

$$\exists v_{n+3} \ldots v_{2n+2} \bigvee_{i=0}^{l} \psi_P(v_0, \ldots, v_{r-1}, \underline{0}, \ldots, \underline{0}, \underline{L}_i, \underline{0}, v_{n+3}, \ldots, v_{2n+2}).$$

(b) We proceed as in (a), noting that

$f(k_0, \ldots, k_{r-1}) = k_r$ iff P, beginning with configuration

$$(0, k_0, \ldots, k_{r-1}, 0, \ldots, 0),$$

after finitely many steps reaches a configuration of the form $(L_i, k_r, m_1, \ldots, m_n)$ with $0 \leq i \leq l$.

Hence the required formula $\varphi(v_0, \ldots, v_{r-1}, v_r)$ can be chosen as

$$\exists v_{n+3} \ldots v_{2n+2} \bigvee_{i=0}^{l} \psi_P(v_0, \ldots, v_{r-1}, \underline{0}, \ldots, \underline{0}, \underline{L}_i, v_r, v_{n+3}, \ldots, v_{2n+2}). \quad \square$$

Relations and functions over $\mathbb{N}$ which can be described by an S_{ar}-formula as in 6.12 are said to be *arithmetical*. Thus 6.12 says that all R-decidable relations and all R-computable functions over $\mathbb{N}$ are arithmetical.

§7. Self-Referential Statements and Gödel's Incompleteness Theorems

In the preceding section we have shown that arithmetic is not R-axiomatizable. Originally Gödel [13] used another method to prove this result. He showed that within sufficiently strong axiom systems there are self-referential formulas, i.e., formulas which make statements about themselves. Such self-referential formulas are the main theme of this section. We shall close by taking up our original objective of this chapter and obtain some important results concerning the limitations of the formal method. With this aim in mind we shall often conduct the arguments on the syntactic level.

In the following we take Φ to be a set of S_{ar}-sentences.

7.1 Definition. (a) A relation $\mathfrak{Q} \subset \mathbb{N}^r$ is said to be *representable in* Φ if there is a formula $\varphi(v_0, \ldots, v_{r-1}) \in L_r^{S_{\mathrm{ar}}}$ such that for all $n_0, \ldots, n_{r-1} \in \mathbb{N}$:

$$\mathfrak{Q} n_0 \ldots n_{r-1} \quad \text{implies } \Phi \vdash \varphi(\underline{n}_0, \ldots, \underline{n}_{r-1}),$$

$$\text{not } \mathfrak{Q} n_0 \ldots n_{r-1} \quad \text{implies } \Phi \vdash \neg\varphi(\underline{n}_0, \ldots, \underline{n}_{r-1}).$$

In this case we say that $\varphi(v_0, \ldots, v_{r-1})$ *represents* $\mathfrak{Q}$ *in* Φ.

(b) A function $F: \mathbb{N}^r \to \mathbb{N}$ is said to be *representable in* Φ if there is a formula $\varphi(v_0, \ldots, v_r) \in L_{r+1}^{S_{\mathrm{ar}}}$ such that for all $n_0, \ldots, n_{r-1}, n_r \in \mathbb{N}$,

$$F(n_0, \ldots, n_{r-1}) = n_r \quad \text{implies } \Phi \vdash \varphi(\underline{n}_0, \ldots, \underline{n}_r);$$

$$F(n_0, \ldots, n_{r-1}) \neq n_r \quad \text{implies } \Phi \vdash \neg\varphi(\underline{n}_0, \ldots, \underline{n}_r);$$

$$\Phi \vdash \exists^{=1} v_r \varphi(\underline{n}_0, \ldots, \underline{n}_{r-1}, v_r).$$

In this case we say that $\varphi(v_0, \ldots, v_r)$ *represents* F *in* Φ.

7.2 Lemma. (a) *If $\Phi \subset \Phi' \subset L_0^{S_{ar}}$ then the relations and functions representable in Φ are also representable in Φ'.*

(b) *If Φ is inconsistent then every relation and every function is representable in Φ.*

(c) *Let Φ be consistent. If Φ is decidable then every relation representable in Φ is decidable and every function representable in Φ is computable.*

All assertions follow directly from definition 7.1. To prove (c), note that the set of formulas derivable from a decidable set Φ is enumerable. □

For $\Phi \subset L_0^{S_{ar}}$ we define

Repr Φ iff all R-decidable relations and all R-computable functions on $\mathbb{N}$ are representable in Φ.

Repr Φ says, in a certain sense, that Φ is rich enough to describe how procedures operate. In the preceding section we have described the execution of programs in $\Phi := \mathrm{Th}(\mathfrak{N})$. Indeed we have

7.3 Proposition. Repr Th($\mathfrak{N}$).

The proof is immediate from 6.12 if one notes that for every S_{ar}-sentence φ,

$$\mathfrak{N} \models \varphi \quad \text{iff} \quad \mathrm{Th}(\mathfrak{N}) \vdash \varphi$$

and

$$\text{not } \mathfrak{N} \models \varphi \quad \text{iff} \quad \mathrm{Th}(\mathfrak{N}) \vdash \neg\varphi.$$ □

A closer analysis shows that the arguments in the previous section which led to a description in Th($\mathfrak{N}$) of the execution of programs, can actually be carried out in Φ_{PA}. Thus one can obtain

7.4 Proposition. Repr Φ_{PA}. □

As an important technical means we assume in the following that an effective coding of the S_{ar}-formulas by natural numbers (a Gödel numbering) is given, and moreover that every number is the Gödel number of some formula. We write n^{φ} for the Gödel number of φ. Conversely, we let φ_n denote the formula with Gödel number n. In this way it is possible to translate statements about formulas into arithmetical statements. For example, a statement about the derivability of a formula φ becomes an arithmetical statement about the Gödel number of φ, and this in turn can be formalized as an S_{ar}-sentence.

The way we shall proceed originates from the liar's paradox, thereby leading, on a formal level, to a clarification of the problems which lie behind this paradox. The paradox of the liar amounts to the fact that the statement

(*) "I am not telling the truth"

can neither be true nor false; for if it were true it would have to be false, and if it were false it would have to be true. Note that (*) makes a statement about itself, and hence is an example of a self-referential statement.

As a first step we consider statements of this kind in general. We show that within a sufficiently rich system, *every* property which is expressible in the system gives rise to a self-referential sentence; more precisely:

7.5 Fixed Point Theorem. *Suppose* Repr Φ. *Then, for every* $\psi \in L_1^{S_{ar}}$, *there is an* S_{ar}*-sentence* φ $(=\varphi_\psi)$ *such that*

$$\Phi \vdash \varphi \leftrightarrow \psi(\underline{n}^\varphi).$$

Intuitively, φ *says "I have the property* ψ*".*

Proof. Let $F: \mathbb{N} \times \mathbb{N} \to \mathbb{N}$ be given by

$$F(n, m) = \begin{cases} n^{\chi(\underline{m})}, & \text{if } n = n^\chi \text{ for some } \chi \in L_1^{S_{ar}} \\ 0, & \text{otherwise.} \end{cases}$$

Clearly, F is computable, and for $\chi \in L_1^{S_{ar}}$ we have

$$F(n^\chi, m) = n^{\chi(\underline{m})}.$$

Since Repr Φ, F can be represented in Φ by a suitable formula $\alpha(v_0, v_1, v_2) \in L_3^{S_{ar}}$. We write x, y, z for v_0, v_1, v_2. For given $\psi \in L_1^{S_{ar}}$ we set

$$\beta := \forall z(\alpha(x, x, z) \to \psi(z)),$$

$$\varphi := \forall z(\alpha(\underline{n}^\beta, \underline{n}^\beta, z) \to \psi(z)).$$

Since $\beta \in L_1^{S_{ar}}$ and $\varphi = \beta\frac{\underline{n}^\beta}{x}$, we have $F(n^\beta, n^\beta) = n^\varphi$ and hence

$$\Phi \vdash \alpha(\underline{n}^\beta, \underline{n}^\beta, \underline{n}^\varphi). \tag{1}$$

We now show in two steps that

$$\Phi \vdash \varphi \leftrightarrow \psi(\underline{n}^\varphi).$$

First, by definition of φ,

$$\Phi \cup \{\varphi\} \vdash \alpha(\underline{n}^\beta, \underline{n}^\beta, \underline{n}^\varphi) \to \psi(\underline{n}^\varphi).$$

By (1), it follows that $\Phi \vdash \varphi \to \psi(\underline{n}^\varphi)$. Since α represents the function F in Φ, we have, on the other hand,

$$\Phi \vdash \exists^{=1} z\, \alpha(\underline{n}^\beta, \underline{n}^\beta, z);$$

thus by (1)

$$\Phi \vdash \forall z(\alpha(\underline{n}^\beta, \underline{n}^\beta, z) \to z \equiv \underline{n}^\varphi),$$

and therefore

$$\Phi \vdash \psi(\underline{n}^\varphi) \to \forall z(\alpha(\underline{n}^\beta, \underline{n}^\beta, z) \to \psi(z)),$$

that is,

$$\Phi \vdash \psi(\underline{n}^\varphi) \to \varphi.$$

□

Let us now turn to systems which are rich enough to contain statements about the "truth" or "falsity" of statements formalizable in the system. The following theorem shows that in such a system one cannot classify all statements as either "true" or "false". Formally, we consider an axiom system Φ; The "true" statements correspond to the sentences in

$$\Phi^{\vdash} := \{\varphi \in L_0^{S_{ar}} \mid \Phi \vdash \varphi\},$$

the "false" statements to the negations of sentences in $\Phi^{\vdash}$. To say that one can speak of "truth" in Φ is to say that $\Phi^{\vdash}$ is representable in Φ (more precisely, that $\{n^{\varphi} \mid \varphi \in \Phi^{\vdash}\}$ is representable in Φ). If Φ satisfies this latter condition then, as we shall show, there is a sentence φ such that neither φ nor $\neg\varphi$ belongs to $\Phi^{\vdash}$.

7.6 Lemma. *Let* Φ *be consistent and suppose* Repr Φ. *If* $\Phi^{\vdash}$ *is representable in* Φ, *then there is an* S_{ar}*-sentence* φ *such that neither* $\Phi \vdash \varphi$ *nor* $\Phi \vdash \neg\varphi$.

PROOF. Suppose $\chi(v_0) \in L_1^{S_{ar}}$ represents the set $\Phi^{\vdash}$ in Φ. Then in particular, for $\alpha \in L_0^{S_{ar}}$,

$$\Phi \vdash \chi(\underline{n}^{\alpha}) \quad \text{iff } \Phi \vdash \alpha.$$

For $\psi = \neg\chi$ we choose, by 7.5, a "fixed point" $\varphi \in L_0^{S_{ar}}$ such that

$$(*) \qquad \Phi \vdash \varphi \leftrightarrow \neg\chi(\underline{n}^{\varphi});$$

φ says intuitively "I am not true".

As in the paradox of the liar, we now obtain that neither $\Phi \vdash \varphi$ nor $\Phi \vdash \neg\varphi$. For if $\Phi \vdash \varphi$ then $\Phi \vdash \chi(\underline{n}^{\varphi})$ and hence by (*), $\Phi \vdash \neg\varphi$, that is, Φ is inconsistent, contrary to our initial assumption. On the other hand, if $\Phi \vdash \neg\varphi$ then by (*), $\Phi \vdash \chi(\underline{n}^{\varphi})$ and therefore $\Phi \vdash \varphi$, and Φ would again be inconsistent. □

Lemma 7.6 has interesting consequences both on the syntactical and semantical levels. In semantical formulations one usually refers to $\Phi^{\models}$ instead of $\Phi^{\vdash}$.

7.7 Tarski's Theorem [26]. (a) *Suppose* Φ *is consistent and* Repr Φ *holds. If* $\Phi^{\models}$ *is representable in* Φ, *then* $\Phi^{\models}$ *is not complete.*
(b) Th($\mathfrak{N}$) *is not representable in* Th($\mathfrak{N}$).

PROOF. (a) Since $\Phi^{\vdash} = \Phi^{\models}$, (a) follows immediately from 7.6.

(b) follows from (a) by setting $\Phi = \text{Th}(\mathfrak{N})$ and noting that $\text{Th}(\mathfrak{N}) = \text{Th}(\mathfrak{N})^{\models}$ is complete. □

Tarski's theorem is of great significance in the study of semantics. Part (a) says that for a sufficiently strong system the following two conditions cannot hold simultaneously:

(1) Every statement in the system is either true or false ("$\Phi^{\models}$ is complete").
(2) Truth is expressible in the system ("$\Phi^{\models}$ is representable in Φ").

The paradox of the liar arises from the tacit assumption that both conditions (1) and (2) hold for everyday language.

Part (b) of Tarski's theorem can be formulated succinctly as "there is no truth definition for arithmetic within arithmetic".

Like Tarski's theorem, Gödel's first incompleteness theorem is also a consequence of lemma 7.6 (cf. 6.10).

7.8 Gödel's First Incompleteness Theorem. *Let Φ be consistent and R-decidable and suppose* Repr Φ. *Then there is an S_{ar}-sentence φ such that neither $\Phi \vdash \varphi$ nor $\Phi \vdash \neg\varphi$.*

PROOF. Suppose that for every S_{ar}-sentence φ, either $\Phi \vdash \varphi$ or $\Phi \vdash \neg\varphi$. Then $\Phi^{\vdash}$ is decidable. Hence, by Repr Φ, $\Phi^{\vdash}$ is representable in Φ in contradiction to 7.6. □

A refinement of the above discussion leads to results concerning the consistency of mathematics. In particular, Gödel's second incompleteness theorem, which we shall derive, shows that the consistency of a sufficiently rich system cannot be proved using only the means available within the system.

Let $\Phi \subset L_0^{S_{ar}}$ be decidable such that Repr Φ. We choose an effective enumeration of all derivations in the sequent calculus associated with S_{ar} and define a relation $H \subset \mathbb{N} \times \mathbb{N}$ by

Hnm iff the mth derivation ends with a sequent of the form $\psi_0 \ldots \psi_{k-1}\varphi_n$, where $\psi_0, \ldots, \psi_{k-1} \in \Phi$ and φ_n is (as before) the nth formula in the Gödel numbering of S_{ar}-formulas.

Since Φ is decidable, so also is H, and clearly,

$$\Phi \vdash \varphi \quad \text{iff there is } m \in \mathbb{N} \text{ such that } Hn^{\varphi}m.$$

As a decidable relation, H can be represented in Φ by a suitable formula $\varphi_H(v_0, v_1) \in L_2^{S_{ar}}$. Again we write x, y for v_0, v_1 and set

$$\mathrm{Der}_{\Phi}(x) := \exists y \varphi_H(x, y).$$

For $\psi = \neg \mathrm{Der}_{\Phi}(x)$ we choose a fixed point $\varphi \in L_0^{S_{ar}}$:

$$(*) \qquad \Phi \vdash \varphi \leftrightarrow \neg \mathrm{Der}_{\Phi}(\underline{n^{\varphi}}).$$

φ says intuitively "I am not provable from Φ".

Then we have

7.9. *If Φ is consistent then not $\Phi \vdash \varphi$.*

PROOF. Suppose $\Phi \vdash \varphi$ holds. Choose m such that $Hn^{\varphi}m$. Then $\Phi \vdash \varphi_H(\underline{n^{\varphi}}, \underline{m})$, and so $\Phi \vdash \mathrm{Der}_{\Phi}(\underline{n^{\varphi}})$. From $(*)$ we have $\Phi \vdash \neg\varphi$, and hence Φ is inconsistent. □

Since Repr Φ, the consistency of Φ is equivalent to "not $\Phi \vdash 0 \equiv 1$". (Note that in case $\Phi \vdash 0 \equiv 1$ every representable set which contains 0 would also contain 1.) The S_{ar}-sentence

$$\mathrm{Consis}_\Phi := \neg \mathrm{Der}_\Phi(\underline{n}^{0 \equiv 1})$$

thus expresses the consistency of Φ. 7.9 may then be formalized as the S_{ar}-sentence

$$\mathrm{Consis}_\Phi \to \neg \mathrm{Der}_\Phi(\underline{n}^{\varphi}).$$

An argument which is in principle simple, though technically rather tedious could now be used to show that the *proof* of 7.9 can be carried out on the basis of Φ, in case $\Phi \supset \Phi_{\mathrm{PA}}$ (and a natural representation $\varphi_{\mathrm{H}}(v_0, v_1)$ of H has been chosen). This means:

(**) $$\Phi \vdash \mathrm{Consis}_\Phi \to \neg \mathrm{Der}_\Phi(\underline{n}^{\varphi}).$$

We then can deduce

7.10 Gödel's Second Incompleteness Theorem. *Let Φ be consistent and R-decidable such that $\Phi_{\mathrm{PA}} \subset \Phi$. Then*

$$\textit{not } \Phi \vdash \mathrm{Consis}_\Phi.$$

Proof. If $\Phi \vdash \mathrm{Consis}_\Phi$ then by (**) $\Phi \vdash \neg \mathrm{Der}_\Phi(\underline{n}^{\varphi})$ also. By (*), $\Phi \vdash \varphi \leftrightarrow \neg \mathrm{Der}_\Phi(\underline{n}^{\varphi})$ and hence $\Phi \vdash \varphi$, in contradiction to 7.9. □

For $\Phi = \Phi_{\mathrm{PA}}$ Gödel's second incompleteness theorem says intuitively that the consistency of Φ_{PA} cannot be proved on the basis of Φ_{PA}. This result shows that Hilbert's program cannot be carried out in its original form. In particular this program aimed at a consistency proof for Φ_{PA} using only finitistic means. The concept "finitistic", though not defined precisely (cf. [18], p. 32), was taken in a very narrow sense; in particular it was required that finitistic proof methods be carried out on the basis of Φ_{PA}.

The above argument can be transferred to other systems where there is a substitute for the natural numbers and where decidable relations and computable functions are representable. In particular, they apply to systems of axioms for set theory such as ZFC. One uses the natural numbers as defined in ZFC. Then one can define an $\{\in\}$-sentence $\mathrm{Consis}_{\mathrm{ZFC}}$, which expresses the consistency of ZFC, to obtain

7.11 Theorem. *If* ZFC *is consistent then not* ZFC $\vdash \mathrm{Consis}_{\mathrm{ZFC}}$. □

Since present-day mathematics can be based on the ZFC axioms, and since "not ZFC $\vdash \mathrm{Consis}_{\mathrm{ZFC}}$" says that the consistency of ZFC cannot be proved using only means available within ZFC, we can formulate 7.11 as follows: If mathematics is consistent, we cannot prove its consistency by mathematical means.

In a similar way also Tarski's theorem and Gödel's first incompleteness theorem can be transferred to axiom systems for set theory. For example, 7.8 would then assert that for every decidable and consistent system of axioms Φ for set theory which contains ZFC, there is an $\{\in\}$-sentence ψ such that neither $\Phi \vdash \psi$ nor $\Phi \vdash \neg\psi$. Intuitively this means that there is no decidable consistent system of axioms for mathematics which, for every mathematical statement, allows us to either prove it or disprove it. In this fact an inherent limitation of the axiomatic method is manifested.

CHAPTER XI

An Algebraic Characterization of Elementary Equivalence

The greater part of our exposition so far has been devoted to the development and investigation of first-order logic. We can justify the dominant rôle assumed by first-order logic in several ways:

(a) First-order logic is in principle sufficient for mathematics.
(b) The intuitive concept of proof and the consequence relation can be adequately described by a formal notion of proof, which is given by means of a calculus.
(c) A number of semantic results such as the compactness theorem or the Löwenheim–Skolem theorem lead to an enrichment of mathematical methods.

However, in contrast to these positive aspects, one also has to take into account that the limited expressive power of first-order language often requires clumsy formulations. In particular, it forces us to make explicit reference to set theory to an extent not usual in mathematical practice. For this reason we were led to seek other systems with greater expressive power but still satisfying conditions (b) and (c). We introduced a number of extensions of first-order logic ($\mathscr{L}_{\mathrm{II}}$, $\mathscr{L}_{\mathrm{II}}^{\mathrm{w}}$, $\mathscr{L}_{\omega_1\omega}$, $\mathscr{L}_Q$) and investigated their semantic properties. In each case we found (cf. IX) that not all the properties mentioned in (c) are available.

In the previous chapter we obtained negative results of a more syntactic nature. For example, we saw that for $\mathscr{L}_{\mathrm{II}}$ and for $\mathscr{L}_{\mathrm{II}}^{\mathrm{w}}$ there is no possibility of adequately describing the notion of proof by means of a calculus; hence in these cases we also have to make concessions concerning (b).

The discussion in the last two chapters will show that these negative results have a deeper reason: Having made precise the concept "logical system" we shall prove in Chapter XII that no logical system with more expressive power than first-order logic can meet the conditions of (b) and (c).

In the present chapter we introduce a useful tool for these investigations. Recall that two structures are elementarily equivalent if they satisfy the same first-order sentences. We now present a purely algebraic characterization of elementary equivalence. This characterization is useful not only for our present purpose but also in other contexts. For example, in many cases it can serve to verify that two given structures $\mathfrak{A}$ and $\mathfrak{B}$ are elementarily equivalent (in a simpler way than by proving directly that $\mathfrak{A}$ and $\mathfrak{B}$ satisfy the same first-order sentences).

§1. Partial Isomorphisms

In this section we provide the concepts we need in order to formulate the algebraic characterization of elementary equivalence. We refer to a fixed symbol set S. The domain of a map p is denoted by $\mathrm{dom}(p)$, its range by $\mathrm{rg}(p)$.

1.1 Definition. Let $\mathfrak{A}$ and $\mathfrak{B}$ be (S-)structures and let p be a map. p is said to be a *partial isomorphism from* $\mathfrak{A}$ *to* $\mathfrak{B}$ if and only if $\mathrm{dom}(p) \subset A$, $\mathrm{rg}(p) \subset B$ and p has the following properties:

(a) p is injective.
(b) p is homomorphic in the following sense:
 (1) For n-ary $P \in S$ and $a_0, \ldots, a_{n-1} \in \mathrm{dom}(p)$,

$$P^{\mathfrak{A}} a_0 \ldots a_{n-1} \quad \text{iff} \quad P^{\mathfrak{B}} p(a_0) \ldots p(a_{n-1}).$$

 (2) For n-ary $f \in S$ and $a_0, \ldots, a_{n-1}, a \in \mathrm{dom}(p)$,

$$f^{\mathfrak{A}}(a_0, \ldots, a_{n-1}) = a \quad \text{iff} \quad f^{\mathfrak{B}}(p(a_0), \ldots, p(a_{n-1})) = p(a).$$

 (3) For $c \in S$ and $a \in \mathrm{dom}(p)$,

$$c^{\mathfrak{A}} = a \quad \text{iff} \quad c^{\mathfrak{B}} = p(a).$$

We write $\mathrm{Part}(\mathfrak{A}, \mathfrak{B})$ for the set of partial isomorphisms from $\mathfrak{A}$ to $\mathfrak{B}$.

1.2 Examples and Comments.

(a) The empty map, i.e., the map with empty domain, is a partial isomorphism from $\mathfrak{A}$ to $\mathfrak{B}$.

(b) The map p with $\mathrm{dom}(p) = \{2, 3\}$ and $p(2) = 2$, $p(3) = 6$ is a partial isomorphism from the additive group $(\mathbb{R}, +, 0)$ of real numbers to the additive group $(\mathbb{Z}, +, 0)$ of integers. On the other hand, the map q with $\mathrm{dom}(q) = \{2, 3\}$ and $q(2) = 1$, $q(3) = 2$ is not a partial isomorphism from $(\mathbb{R}, +, 0)$ to $(\mathbb{Z}, +, 0)$, because $2 + 2 \neq 3$ but $q(2) + q(2) = q(3)$.

(c) If S is relational, that is, if S contains only relation symbols, then for $a_0, \ldots, a_{r-1} \in A$ and $b_0, \ldots, b_{r-1} \in B$ the following two statements are equivalent:

(*) By setting

$$p(a_i) := b_i \qquad (i < r),$$

a partial isomorphism from $\mathfrak{A}$ to $\mathfrak{B}$ is determined (where $\operatorname{dom}(p) = \{a_0, \ldots, a_{r-1}\}$ and $\operatorname{rg}(p) = \{b_0, \ldots, b_{r-1}\}$).

(**) For every atomic formula $\psi \in L_r^S$,

$$\mathfrak{A} \models \psi[a_0, \ldots, a_{r-1}] \quad \text{iff } \mathfrak{B} \models \psi[b_0, \ldots, b_{r-1}].$$

Proof. First we note that for $i, j < r$

$$(1) \qquad \begin{aligned} a_i = a_j &\quad \text{iff } \mathfrak{A} \models v_i \equiv v_j[a_0, \ldots, a_{r-1}], \\ b_i = b_j &\quad \text{iff } \mathfrak{B} \models v_i \equiv v_j[b_0, \ldots, b_{r-1}], \end{aligned}$$

and that for n-ary $P \in S$ and $i_0, \ldots, i_{n-1} < r$

$$(2) \qquad \begin{aligned} P^{\mathfrak{A}} a_{i_0} \ldots a_{i_{n-1}} &\quad \text{iff } \mathfrak{A} \models P v_{i_0} \ldots v_{i_{n-1}}[a_0, \ldots, a_{r-1}], \\ P^{\mathfrak{B}} b_{i_0} \ldots b_{i_{n-1}} &\quad \text{iff } \mathfrak{B} \models P v_{i_0} \ldots v_{i_{n-1}}[b_0, \ldots, b_{r-1}]. \end{aligned}$$

Now, if (**) holds, then by (1) and the fact that

$$\mathfrak{A} \models v_i \equiv v_j[a_0, \ldots, a_{r-1}] \quad \text{iff } \mathfrak{B} \models v_i \equiv v_j[b_0, \ldots, b_{r-1}],$$

p (as given in (*)) is well-defined and injective. Since

$$\mathfrak{A} \models P v_{i_0} \ldots v_{i_{n-1}}[a_0, \ldots, a_{r-1}] \quad \text{iff } \mathfrak{B} \models P v_{i_0} \ldots v_{i_{n-1}}[b_0, \ldots, b_{r-1}],$$

and by (2), p is also homomorphic.

Similarly one can use (1) and (2) to deduce (**) from (*). ☐

(d) Note that the equivalence in (c) is no longer true if S contains function symbols or constants. For example, for the partial isomorphism p in (b)

$$\text{not } (\mathbb{R}, +, 0) \models v_0 + (v_0 + v_0) \equiv v_1[2, 3],$$

but on the other hand

$$(\mathbb{Z}, +, 0) \models v_0 + (v_0 + v_0) \equiv v_1[p(2), p(3)].$$

(e) The following example shows that even for relational S a partial isomorphism does not in general preserve the validity of formulas with quantifiers.

Let $S = \{<\}$ and let q_0 be the partial isomorphism from $(\mathbb{R}, <)$ to $(\mathbb{Z}, <)$ such that $\operatorname{dom}(q_0) = \{2, 3\}$ and $q_0(2) = 3$, $q_0(3) = 4$. Then

$$(\mathbb{R}, <) \models \exists v_2(v_0 < v_2 \land v_2 < v_1)\,[2, 3]$$

but

$$\text{not } (\mathbb{Z}, <) \models \exists v_2(v_0 < v_2 \land v_2 < v_1)\,[q_0(2), q_0(3)].$$

If p is a partial isomorphism from $(\mathbb{R}, <)$ to $(\mathbb{Z}, <)$ such that $\operatorname{dom}(p) = \{a, b\}$ and $a < b$, then we always have

$$(\mathbb{R}, <) \models \exists v_2(v_0 < v_2 \wedge v_2 < v_1)\,[a, b],$$

since, for example,

$$(\mathbb{R}, <) \models v_0 < v_2 \wedge v_2 < v_1\left[a, b, \frac{a+b}{2}\right].$$

In this case the validity of

$$(+) \qquad (\mathbb{Z}, <) \models \exists v_2(v_0 < v_2 \wedge v_2 < v_1)[p(a), p(b)]$$

is equivalent to the existence of a partial isomorphism q from $(\mathbb{R}, <)$ to $(\mathbb{Z}, <)$ which extends p and has $(a + b)/2$ in its domain. For, if such a q exists, then $(+)$ holds, since

$$(\mathbb{Z}, <) \models v_0 < v_2 \wedge v_2 < v_1\left[q(a), q(b), q\left(\frac{a+b}{2}\right)\right];$$

conversely, if $(+)$ is satisfied and, say,

$$(\mathbb{Z}, <) \models v_0 < v_2 \wedge v_2 < v_1[p(a), p(b), d],$$

then the extension q of p with $\operatorname{dom}(q) = \{a, b, (a + b)/2\}$ and $q((a + b)/2) = d$ is such a partial isomorphism.

This argument indicates that the truth of formulas with quantifiers is preserved under partial isomorphisms provided that these admit certain extensions. It embodies the basic idea behind the algebraic characterization of elementary equivalence: The elementary equivalence of structures amounts to the existence of extensions of certain partial isomorphisms.

In the following we identify a map p with its graph $\{(a, p(a)) \mid a \in \operatorname{dom}(p)\}$. Then $p \subset q$ means that q is an extension of p.

1.3 Definition. $\mathfrak{A}$ and $\mathfrak{B}$ are said to be *finitely isomorphic*, written $\mathfrak{A} \cong_f \mathfrak{B}$, iff there is a sequence $(I_n)_{n\in\mathbb{N}}$ with the following properties:

(a) Every I_n is a nonempty set of partial isomorphisms from $\mathfrak{A}$ to $\mathfrak{B}$.
(b) (Forth-property) For every $p \in I_{n+1}$ and $a \in A$ there is $q \in I_n$ such that $q \supset p$ and $a \in \operatorname{dom}(q)$.
(c) (Back-property) For every $p \in I_{n+1}$ and $b \in B$ there is $q \in I_n$ such that $q \supset p$ and $b \in \operatorname{rg}(q)$.

Informally we can express (b) and (c) as follows: partial isomorphisms in I_{n+1} can be extended $(n + 1)$ times; the corresponding extensions lie in $I_n, I_{n-1}, \ldots, I_1$, and I_0, respectively. If $(I_n)_{n\in\mathbb{N}}$ has the properties (a), (b), and (c), we write $(I_n)_{n\in\mathbb{N}}\colon \mathfrak{A} \cong_f \mathfrak{B}$.

1.4 Definition. $\mathfrak{A}$ and $\mathfrak{B}$ are said to be *partially isomorphic*, written $\mathfrak{A} \cong_p \mathfrak{B}$, iff there is a set I such that

(a) I is a nonempty set of partial isomorphisms from $\mathfrak{A}$ to $\mathfrak{B}$.
(b) (Forth-property) For every $p \in I$ and $a \in A$ there is $q \in I$ such that $q \supset p$ and $a \in \text{dom}(q)$.
(c) (Back-property) For every $p \in I$ and $b \in B$ there is $q \in I$ such that $q \supset p$ and $b \in \text{rg}(q)$.

Thus the conditions (a), (b), and (c) amount to $(I)_{n \in \mathbb{N}}: \mathfrak{A} \cong_f \mathfrak{B}$ for the constant sequence $(I)_{n \in \mathbb{N}}$.

If (a), (b), and (c) are satisfied for I we write $I: \mathfrak{A} \cong_p \mathfrak{B}$.

The following lemma lists the relations between the various notions of isomorphism.

1.5 Lemma. (a) *If $\mathfrak{A} \cong \mathfrak{B}$, then $\mathfrak{A} \cong_p \mathfrak{B}$.*
(b) *If $\mathfrak{A} \cong_p \mathfrak{B}$ then $\mathfrak{A} \cong_f \mathfrak{B}$.*
(c) *If $\mathfrak{A} \cong_f \mathfrak{B}$ and A is finite then $\mathfrak{A} \cong \mathfrak{B}$.*
(d) *If $\mathfrak{A} \cong_p \mathfrak{B}$ and A and B are at most countable then $\mathfrak{A} \cong \mathfrak{B}$.*

PROOF. (a) If $\pi: \mathfrak{A} \cong \mathfrak{B}$ then $I: \mathfrak{A} \cong_p \mathfrak{B}$ for $I = \{\pi\}$.

(b) If $I: \mathfrak{A} \cong_p \mathfrak{B}$ then $(I)_{n \in \mathbb{N}}: \mathfrak{A} \cong_f \mathfrak{B}$.

(c) Suppose $(I_n)_{n \in \mathbb{N}}: \mathfrak{A} \cong_f \mathfrak{B}$, and suppose A has exactly r elements, $A = \{a_0, \ldots, a_{r-1}\}$. We choose $p \in I_{r+1}$. If we suitably apply the forth-property r times we obtain a $q \in I_1$ such that $a_0, \ldots, a_{r-1} \in \text{dom}(q)$, i.e., $\text{dom}(q) = A$. If $\text{rg}(q) \neq B$ and $b \in B - \text{rg}(q)$, then by the back-property there would be an extension q' of q in I_0 such that $b \in \text{rg}(q')$. Since $\text{dom}(q) = A$, this is not possible. Therefore $\text{rg}(q) = B$ and thus $q: \mathfrak{A} \cong \mathfrak{B}$.

(d) Suppose $I: \mathfrak{A} \cong_p \mathfrak{B}$, $A = \{a_0, a_1, \ldots\}$ and $B = \{b_0, b_1, \ldots\}$. Starting from an arbitrary $p_0 \in I$, by repeated application of the back- and forth-properties, we obtain extensions $p_1, p_2, \ldots$ in I such that $a_0 \in \text{dom}(p_1)$, $b_0 \in \text{rg}(p_2)$, $a_1 \in \text{dom}(p_3)$, $b_1 \in \text{rg}(p_4)$, $\ldots$, that is

(1) $p_n \subset p_{n+1}$;
(2) if n is odd, say $n = 2r + 1$, then $a_r \in \text{dom}(p_n)$;
(3) if n is even, say $n = 2r + 2$, then $b_r \in \text{rg}(p_n)$.

By (1), $p := \bigcup_{n \in \mathbb{N}} p_n$ is a partial isomorphism from $\mathfrak{A}$ to $\mathfrak{B}$. As $\text{dom}(p) = A$ (by (2)) and $\text{rg}(p) = B$ (by (3)), we have $p: \mathfrak{A} \cong \mathfrak{B}$. □

Part (d) of 1.5 is an abstract version of the following theorem of Cantor:

1.6 Theorem. *Any two countable dense orderings (without endpoints) are isomorphic.*

Here a *dense ordering* is a $\{<\}$-structure which is a model of Φ_{dord}, where Φ_{dord} contains the ordering axioms together with the following sentences:

$$\forall x \, \forall y (x < y \rightarrow \exists z (x < z \wedge z < y)),$$

$$\forall x \, \exists y \; x < y, \qquad \forall x \, \exists y \; y < x.$$

$(\mathbb{R}, <)$ and $(\mathbb{Q}, <)$ are dense orderings. By contrast, $(\mathbb{Z}, <)$ is not a dense ordering.

Cantor's theorem follows from 1.5(d) and

1.7 Lemma. *If* $\mathfrak{A} = (A, <^A)$ *and* $\mathfrak{B} = (B, <^B)$ *are dense orderings, then* $I: \mathfrak{A} \cong_p \mathfrak{B}$ *for* $I = \{p \mid p \in \text{Part}(\mathfrak{A}, \mathfrak{B}), \text{dom}(p)$ *finite*$\}$.

PROOF. Since $p = \emptyset$ is in I, I is not empty. I satisfies the forth-property. For, if $p \in I$, $\text{dom}(p) = \{a_0, \ldots, a_{n-1}\}$, and $a \in A$, then because $\mathfrak{B}$ is dense there is an element $b \in B$ which is related to $p(a_0), \ldots, p(a_{n-1})$ in the ordering $\mathfrak{B}$ in the same manner as a is related to $a_0, \ldots, a_{n-1}$ in the ordering $\mathfrak{A}$. Then $q := p \cup \{(a, b)\}$ is an extension of p which is defined for a and lies in I. The back-property follows analogously, using the fact that $\mathfrak{A}$ is dense. □

1.8 Example. Suppose $S = \{\underline{\sigma}, 0\}$ and let Φ_σ consist of the "successor axioms"

$$\forall x(\neg x \equiv 0 \leftrightarrow \exists y \underline{\sigma} y \equiv x),$$

$$\forall x \forall y(\underline{\sigma} x \equiv \underline{\sigma} y \rightarrow x \equiv y),$$

and for every $m \geq 1$:

$$\forall x \neg \underbrace{\underline{\sigma} \ldots \underline{\sigma}}_{m\text{-times}} x \equiv x.$$

The structure $\mathfrak{N}_\sigma$ (cf. III.7.3(2)) is a model of Φ_σ. We show that any two models of Φ_σ are finitely isomorphic. For a model $\mathfrak{A}$ of Φ_σ and for $a \in A$, let

$$a^{(m)} := \underbrace{\underline{\sigma}^A \ldots \underline{\sigma}^A}_{m\text{-times}}(a).$$

For every $n \in \mathbb{N}$ we define a "distance function" d_n on $A \times A$ by

$$d_n(a, a') := \begin{cases} m & \text{if } a^{(m)} = a' \text{ and } m \leq 2^n, \\ -m & \text{if } a'^{(m)} = a \text{ and } m \leq 2^n, \\ \infty & \text{otherwise.} \end{cases}$$

Now suppose $\mathfrak{A}$ and $\mathfrak{B}$ are models of Φ_σ. We show that $(I_n)_{n \in \mathbb{N}}: \mathfrak{A} \cong_f \mathfrak{B}$, where

$$I_n := \{p \in \text{Part}(\mathfrak{A}, \mathfrak{B}) \mid \text{dom}(p) \text{ finite}, 0^A \in \text{dom}(p),$$
$$\text{and for all } a, a' \in \text{dom}(p), d_n(a, a') = d_n(p(a), p(a'))\}.$$

Thus a partial isomorphism in I_n preserves the "d_n-distances". First we have $I_n \neq \emptyset$ since $\{(0^A, 0^B)\} \in I_n$. We sketch a proof of the forth-property for $(I_n)_{n \in \mathbb{N}}$ (the back-property can be proved analogously). Suppose $p \in I_{n+1}$ and $a \in A$. We distinguish two cases, depending on whether or not the following condition (∗) is satisfied.

(∗) There is an $a' \in \text{dom}(p)$ such that $|d_n(a', a)| \leq 2^n$.

If $(*)$ holds there is exactly one $b \in B$ for which $p \cup \{(a, b)\}$ preserves the d_n-distance (since $p \in I_{n+1}$); if $(*)$ does not hold we choose an arbitrary element b such that $d_n(p(a'), b) = \infty$ for all $a' \in \mathrm{dom}(p)$ (such an element must exist since every model of Φ_σ is infinite). In any case it is easy to show that $q := p \cup \{(a, b)\} \in I_n$. □

1.9 Exercise. Let $S = \emptyset$. Show that any two infinite S-structures are partially isomorphic.

1.10 Exercise. (a) Give an example of structures which are partially isomorphic but not isomorphic.

(b) Give an example of structures which are finitely isomorphic but not partially isomorphic.

1.11 Exercise. Give an uncountable model of the system of axioms Φ_σ in 1.8.

§2. Fraissé's Theorem

Using the concepts introduced in §1, we now formulate the main result of this chapter.

2.1 Fraissé's Theorem. *Let S be a finite symbol set and $\mathfrak{A}$, $\mathfrak{B}$ S-structures. Then*

$$\mathfrak{A} \equiv \mathfrak{B} \quad \textit{iff} \quad \mathfrak{A} \cong_f \mathfrak{B}.$$

Note that Fraissé's theorem provides us with a characterization of elementary equivalence which does not refer to first-order language.

Before proving the theorem (in the next section) we give several examples showing how it can be used to check the elementary equivalence of structures and the completeness of theories.

2.2 Proposition. (a) *Any two dense orderings are elementarily equivalent. In particular, $(\mathbb{R}, <) \equiv (\mathbb{Q}, <)$.*

(b) *Any two $\{\sigma, 0\}$-structures satisfying the axioms in 1.8 are elementarily equivalent.*

Proof. (a) follows from 2.1, since every two dense orderings are partially isomorphic, and thus also finitely isomorphic; (b) follows analogously by means of 1.8. □

For some applications we need the following simple criterion for the completeness of theories.

2.3 Lemma. *For a theory $T \subset L_0^S$ the following are equivalent:*

(a) *T is complete, i.e., for every S-sentence φ either $\varphi \in T$ or $\neg\varphi \in T$.*

(b) *Any two models of T are elementarily equivalent.*

PROOF. Suppose first that (a) holds, and let $\mathfrak{A}$, $\mathfrak{B}$ be models of T. For any S-sentence φ either $\varphi \in T$ or $\neg\varphi \in T$. If $\varphi \in T$ then $\mathfrak{A} \models \varphi$ and $\mathfrak{B} \models \varphi$; if $\neg\varphi \in T$ then $\mathfrak{A} \models \neg\varphi$ and $\mathfrak{B} \models \neg\varphi$. Thus $\mathfrak{A} \models \varphi$ iff $\mathfrak{B} \models \varphi$.

Conversely, let φ be an S-sentence and suppose $\varphi \notin T$. Since T is a theory, $T \models \varphi$ does not hold, and therefore there is a model $\mathfrak{A}$ of $T \cup \{\neg\varphi\}$. By (b) every model of T is elementarily equivalent to $\mathfrak{A}$, and thus is a model of $\neg\varphi$. Hence $T \models \neg\varphi$ and therefore $\neg\varphi \in T$. □

From 2.2, with the aid of 2.3 and X.6.5 we obtain

2.4 Proposition. (a) *The theory* $\Phi_{\text{dord}}^{\models}$ *of dense orderings is complete and R-decidable.* (*Thus, for example,* $\Phi_{\text{dord}}^{\models} = \text{Th}(\mathbb{R}, <)$.)

(b) *The theory* $\Phi_{\sigma}^{\models}$ *of "successor structures" is complete and R-decidable.* (*Thus for example,* $\Phi_{\sigma}^{\models} = \text{Th}(\mathbb{N}, \sigma)$.) □

Preparatory to the proof of Fraissé's theorem we show that we can restrict ourselves to relational symbol sets.

Let S be an arbitrary symbol set. As on p. 120 we choose, for each n-ary $f \in S$, a new $(n+1)$-ary relation symbol F and, for each $c \in S$, a new unary relation symbol C. Let S^r consist of the relation symbols from S together with the new relation symbols. S^r is relational. For an S-structure $\mathfrak{A}$, let $\mathfrak{A}^r$ be the S^r-structure obtained from $\mathfrak{A}$, replacing functions and constants by their graphs (as in VIII.1).

When defining partial isomorphisms we treated functions and constants in such a way (cf. 1.1(b)) that

$$\text{Part}(\mathfrak{A}, \mathfrak{B}) = \text{Part}(\mathfrak{A}^r, \mathfrak{B}^r).$$

From this we obtain

$$(*) \qquad \mathfrak{A} \cong_f \mathfrak{B} \quad \text{iff} \quad \mathfrak{A}^r \cong_f \mathfrak{B}^r.$$

In VIII.1.7 we showed that

$$(**) \qquad \mathfrak{A} \equiv \mathfrak{B} \quad \text{iff} \quad \mathfrak{A}^r \equiv \mathfrak{B}^r.$$

Thus, in proving Fraissé's theorem, we can restrict to *relational* symbol sets. For, if $\mathfrak{A}$ and $\mathfrak{B}$ are given, it follows from

$$\mathfrak{A}^r \equiv \mathfrak{B}^r \quad \text{iff} \quad \mathfrak{A}^r \cong_f \mathfrak{B}^r$$

by $(*)$ and $(**)$ that

$$\mathfrak{A} \equiv \mathfrak{B} \quad \text{iff} \quad \mathfrak{A} \cong_f \mathfrak{B}.$$

2.5 Exercise. Show that for $S = \varnothing$ the theory of infinite sets, $\{\varphi_{\geq n} \mid n \geq 2\}^{\models}$, is complete and R-decidable.

2.6 Exercise. Let $S = \{P_n | n \in \mathbb{N}\}$ be a set of unary relation symbols. Define the S-structures $\mathfrak{A}$ and $\mathfrak{B}$ as follows: $A := \mathbb{N}$, $B := \mathbb{N} \cup \{\infty\}$, $P_n^{\mathfrak{A}} := \{m | m \in \mathbb{N}, m \geq n\}$, $P_n^{\mathfrak{B}} := \{m | m \in \mathbb{N}, m \geq n\} \cup \{\infty\}$. Show that $\mathfrak{A} \equiv \mathfrak{B}$ but not $\mathfrak{A} \cong_f \mathfrak{B}$. (Thus Fraissé's theorem is in general not true for infinite symbol sets. Note, on the other hand, that for arbitrary S and S-structures $\mathfrak{A}$, $\mathfrak{B}$ we have $\mathfrak{A} \equiv \mathfrak{B}$ iff for any finite $S_0 \subset S$, $\mathfrak{A} \upharpoonright S_0 \equiv \mathfrak{B} \upharpoonright S_0$, that is, $\mathfrak{A} \upharpoonright S_0 \cong_f \mathfrak{B} \upharpoonright S_0$.)

§3. Proof of Fraissé's Theorem

As a measure of the complexity of formulas we define the *quantifier rank* of a formula φ to be the maximum number of nested quantifiers occurring in it:

$$\mathrm{qr}(\varphi) := 0, \quad \text{if } \varphi \text{ is atomic};$$

$$\mathrm{qr}(\neg\varphi) := \mathrm{qr}(\varphi); \qquad \mathrm{qr}(\varphi \vee \psi) := \max\{\mathrm{qr}(\varphi), \mathrm{qr}(\psi)\};$$

$$\mathrm{qr}(\exists x\varphi) := \mathrm{qr}(\varphi) + 1.$$

For example, the formula $\neg\exists x(\forall y\, Rxz \wedge Qy) \wedge \forall z Qz$ has quantifier rank 2. The formulas of quantifier rank zero are the quantifier-free formulas.

In the sequel let S be a fixed finite relational symbol set. One half of Fraissé's theorem amounts to

3.1. *If* $\mathfrak{A} \cong_f \mathfrak{B}$ *then* $\mathfrak{A} \equiv \mathfrak{B}$.

In order to prove 3.1 we must show for every S-sentence φ that

$$\mathfrak{A} \models \varphi \quad \text{iff} \quad \mathfrak{B} \models \varphi.$$

We obtain this by applying the following lemma, taking $r = 0$, $n = \mathrm{qr}(\varphi)$, and an arbitrary $p \in I_n$ (note that $I_n \neq \emptyset$).

3.2 Lemma. *Let* $(I_n)_{n \in \mathbb{N}}: \mathfrak{A} \cong_f \mathfrak{B}$. *Then for every formula* φ:

(∗) *If* $\varphi \in L_r^S$, $\mathrm{qr}(\varphi) \leq n$, $p \in I_n$, *and* $a_0, \ldots, a_{r-1} \in \mathrm{dom}(p)$, *then*

$$\mathfrak{A} \models \varphi[a_0, \ldots, a_{r-1}] \textit{ iff } \mathfrak{B} \models \varphi[p(a_0), \ldots, p(a_{r-1})].$$

Informally, 3.2 says that partial isomorphisms from I_n preserve formulas of quantifier rank $\leq n$. It makes precise the idea discussed in 1.2(e) that formulas with quantifiers are preserved under partial isomorphisms provided these isomorphisms admit certain extensions.

Proof of 3.2. We show (∗) by induction on formulas φ. Suppose $\varphi \in L_r^S$, $\mathrm{qr}(\varphi) \leq n$, $p \in I_n$, and $a_0, \ldots, a_{r-1} \in \mathrm{dom}(p)$.

(i) For atomic φ the result was proved in 1.2(c).

(ii) If $\varphi = \neg\psi$ then

$$\begin{aligned}\mathfrak{A} \models \varphi[a_0, \ldots, a_{r-1}] \quad &\text{iff not } \mathfrak{A} \models \psi[a_0, \ldots, a_{r-1}] \\ &\text{iff not } \mathfrak{B} \models \psi[p(a_0), \ldots, p(a_{r-1})] \\ &\qquad\qquad\text{(by induction hypothesis)} \\ &\text{iff } \mathfrak{B} \models \varphi[p(a_0), \ldots, p(a_{r-1})].\end{aligned}$$

(iii) For $\varphi = \psi_0 \vee \psi_1$ the argument is analogous.

(iv) Suppose $\varphi = \exists x\psi$. Since $\varphi \in L_r^S$, v_r does not occur free in φ. Thus $\models \exists x\psi \leftrightarrow \exists v_r \, \psi\frac{v_r}{x}$, and therefore we may assume that $x = v_r$. Because $\mathrm{qr}(\varphi) = \mathrm{qr}(\exists x\psi) \leq n$, we have $\mathrm{qr}(\psi) \leq n - 1$. The claim for φ is now obtained from the following chain of equivalent statements:

(a) $\mathfrak{A} \models \varphi[a_0, \ldots, a_{r-1}]$.

(b) There is $a \in A$ such that $\mathfrak{A} \models \psi[a_0, \ldots, a_{r-1}, a]$.

(c) There is $a \in A$ and $q \in I_{n-1}$ such that $q \supset p$, $a \in \mathrm{dom}(q)$, and $\mathfrak{A} \models \psi[a_0, \ldots, a_{r-1}, a]$.

(d) There is $a \in A$ and $q \in I_{n-1}$ such that $q \supset p$, $a \in \mathrm{dom}(q)$, and $\mathfrak{B} \models \psi[p(a_0), \ldots, p(a_{r-1}), q(a)]$.

(e) There is $b \in B$ and $q \in I_{n-1}$ such that $q \supset p$, $b \in \mathrm{rg}(q)$, and

$$\mathfrak{B} \models \psi[p(a_0), \ldots, p(a_{r-1}), b].$$

(f) There is $b \in B$ such that $\mathfrak{B} \models \psi[p(a_0), \ldots, p(a_{r-1}), b]$.

(g) $\mathfrak{B} \models \varphi[p(a_0), \ldots, p(a_{r-1})]$.

To prove the equivalence of (e) and (f) and of (b) and (c), respectively, one uses the back- and forth-properties of the sequence $(I_n)_{n \in \mathbb{N}}$. The equivalence of (c) and (d) follows from the induction hypothesis. □

From the foregoing proof we extract another result needed in the next chapter. Two structures $\mathfrak{A}$ and $\mathfrak{B}$ are *m-isomorphic* (written: $\mathfrak{A} \cong_m \mathfrak{B}$) if there is a sequence $I_0, \ldots, I_m$ of nonempty sets of partial isomorphisms from $\mathfrak{A}$ to $\mathfrak{B}$ with the back- and forth-properties, i.e.,

> for $n + 1 \leq m$, $p \in I_{n+1}$ and $a \in A$ (resp. $b \in B$), there is $q \in I_n$ such that $q \supset p$ and $a \in \mathrm{dom}(q)$ (resp. $b \in \mathrm{rg}(q)$).

In this case we write $(I_n)_{n \leq m}\colon \mathfrak{A} \cong_m \mathfrak{B}$.

Since the proof of 3.2 shows that partial isomorphisms in I_m preserve formulas of quantifier rank $\leq m$, we have

3.3 Corollary. *If $\mathfrak{A} \cong_m \mathfrak{B}$ then $\mathfrak{A}$ and $\mathfrak{B}$ satisfy the same sentences of quantifier rank $\leq m$.* □

The following considerations are needed for the converse of 3.1.

Given a set Φ of S-formulas we write $\langle\Phi\rangle$, as in VIII.3, to denote the smallest subset of L^S which contains Φ and is closed under propositional connectives (i.e., which contains together with ψ and χ also $\neg\psi$ and $(\psi \vee \chi)$).

It is easy to verify:

(1) If every formula in Φ_1 is logically equivalent to a formula in Φ_2, then every formula in $\langle\Phi_1\rangle$ is logically equivalent to a formula in $\langle\Phi_2\rangle$.

By induction on φ one can show:

(2) If $\varphi \in L_r^S$ and $\mathrm{qr}(\varphi) \leq n + 1$ then

$$\varphi \in \langle\{\psi \in L_r^S \mid \mathrm{qr}(\psi) \leq n\} \cup \{\exists x\, \psi \in L_r^S \mid \mathrm{qr}(\psi) \leq n\}\rangle.$$

From (1) and (2) we obtain

3.4 Lemma. *Let $n \in \mathbb{N}$. Then for every $r \in \mathbb{N}$ there are, up to logical equivalence, only finitely many formulas in L_r^S of quantifier rank $\leq n$.*

Proof. By induction on n.

$n = 0$: Since S was assumed to be finite and relational, there are only finitely many atomic formulas in L_r^S. By VIII.3.2 there are, up to logical equivalence, only finitely many quantifier-free formulas in L_r^S.

Induction step: Let $r \in \mathbb{N}$. By induction hypothesis there are formulas

$$\psi_0, \ldots, \psi_{k-1} \in L_r^S \text{ of quantifier rank } \leq n$$

and formulas

$$\chi_0, \ldots, \chi_{h-1} \in L_{r+1}^S \text{ of quantifier rank } \leq n$$

such that every formula in L_r^S (resp. L_{r+1}^S) of quantifier rank $\leq n$ is logically equivalent to some ψ_i (resp. χ_i). We show:

(*) Every formula in

$$\Phi_1 := \{\psi \in L_r^S \mid \mathrm{qr}(\psi) \leq n\} \cup \{\exists x\, \psi \in L_r^S \mid \mathrm{qr}(\psi) \leq n\}$$

is logically equivalent to a formula in

$$\Phi_2 := \{\psi_0, \ldots, \psi_{k-1}\} \cup \{\exists v_r\, \chi_0, \ldots, \exists v_r\, \chi_{h-1}\}.$$

From this we obtain the claim in the induction step as follows: Every formula in L_r^S of quantifier rank $\leq n + 1$ is in $\langle\Phi_1\rangle$ (cf. (2)) and is therefore, by (*) and (1), logically equivalent to a formula in $\langle\Phi_2\rangle$. But by VIII.3.2 $\langle\Phi_2\rangle$ contains only finitely many formulas which are pairwise logically non-equivalent.

Proof of (*): By choice of the ψ_i every $\psi \in L_r^S$ with $\mathrm{qr}(\psi) \leq n$ is logically equivalent to some ψ_i. If $\exists x \psi \in L_r^S$ and $\mathrm{qr}(\psi) \leq n$, then $\models \exists x \psi \leftrightarrow \exists v_r \psi \frac{v_r}{x}$.

The formula $\psi\frac{v_r}{x}$ is in L^S_{r+1} and has quantifier rank $\leq n$, hence is logically equivalent to some χ_i; thus $\exists x\,\psi$ is equivalent to $\exists v_r\,\chi_i$. □

We conclude the proof of Fraissé's theorem by showing

3.5 Lemma. *If* $\mathfrak{A} \equiv \mathfrak{B}$ *then* $\mathfrak{A} \cong_f \mathfrak{B}$.

PROOF. Suppose $\mathfrak{A} \equiv \mathfrak{B}$. For $n \in \mathbb{N}$ we define I_n as follows (cf. (*) in 3.2):

$p \in I_n$ iff $p \in \mathrm{Part}(\mathfrak{A}, \mathfrak{B})$ and there are $r \in \mathbb{N}$ and $a_0, \ldots, a_{r-1} \in A$ with $\mathrm{dom}(p) = \{a_0, \ldots, a_{r-1}\}$ such that for all $\varphi \in L^S_r$ with $\mathrm{qr}(\varphi) \leq n$

$$\mathfrak{A} \models \varphi[a_0, \ldots, a_{r-1}] \quad \text{iff } \mathfrak{B} \models \varphi[p(a_0), \ldots, p(a_{r-1})].$$

We show that $(I_n)_{n\in\mathbb{N}}\colon \mathfrak{A} \cong_f \mathfrak{B}$.

$(I_n)_{n\in\mathbb{N}}$ has the forth-property. For, suppose that $p \in I_{n+1}$, $a \in A$ and $\mathrm{dom}(p) = \{a_0, \ldots, a_{r-1}\}$. By 3.4 we can pick finitely many formulas $\psi_0, \ldots, \psi_s \in L^S_{r+1}$ of quantifier rank $\leq n$ such that every formula in L^S_{r+1} of quantifier rank $\leq n$ is logically equivalent to some ψ_i.

For $0 \leq i \leq s$ we let

$$\varphi_i := \begin{cases} \psi_i, & \text{if } \mathfrak{A} \models \psi_i[a_0, \ldots, a_{r-1}, a], \\ \neg\psi_i, & \text{if } \mathfrak{A} \models \neg\psi_i[a_0, \ldots, a_{r-1}, a]. \end{cases}$$

Then $\mathfrak{A} \models \exists v_r(\varphi_0 \wedge \cdots \wedge \varphi_s)[a_0, \ldots, a_{r-1}]$. Since

$$\mathrm{qr}(\exists v_r(\varphi_0 \wedge \cdots \wedge \varphi_s)) \leq n + 1$$

and $p \in I_{n+1}$, it follows that

$$\mathfrak{B} \models \exists v_r(\varphi_0 \wedge \cdots \wedge \varphi_s)[p(a_0), \ldots, p(a_{r-1})],$$

say $\mathfrak{B} \models \varphi_0 \wedge \cdots \wedge \varphi_s[p(a_0), \ldots, p(a_{r-1}), b]$. Then in $\mathfrak{A}$ and $\mathfrak{B}$ the elements $a_0, \ldots, a_{r-1}, a$ and $p(a_0), \ldots, p(a_{r-1}), b$ respectively, satisfy the same formulas among the ψ_i and therefore satisfy the same formulas of quantifier rank $\leq n$. Thus $p = q \cup \{(a, b)\}$ is a partial isomorphism which extends p (cf. 1.2(c)), has a in its domain, and lies in I_n.

The back-property is proved analogously.

Finally, every I_n is nonempty: Since $\mathfrak{A} \equiv \mathfrak{B}$ the same sentences of quantifier rank $\leq n$ hold in $\mathfrak{A}$ as in $\mathfrak{B}$, and therefore $p = \emptyset$ lies in I_n. □

If $\mathfrak{A}$ and $\mathfrak{B}$ satisfy the same sentences of quantifier rank $\leq m$, the last argument in the preceding proof shows that $p = \emptyset$ is an element of I_0, $I_1, \ldots, I_m$. Summarizing, we have

3.6 Lemma. *If* $\mathfrak{A}$ *and* $\mathfrak{B}$ *satisfy the same sentences of quantifier rank* $\leq m$ *then* $\mathfrak{A} \cong_m \mathfrak{B}$. □

Assertions 3.3 and 3.6 yield

3.7 Theorem. *Let S be finite and relational. For S-structures $\mathfrak{A}$ and $\mathfrak{B}$, the following are equivalent:*

(a) $\mathfrak{A} \cong_m \mathfrak{B}$.
(b) $\mathfrak{A}$ *and* $\mathfrak{B}$ *satisfy the same sentences of quantifier rank* $\leq m$. □

3.8 Exercise. Suppose $S = \{P_0, \ldots, P_{r-1}\}$ with unary relation symbols P_i. Show that for every S-structure $\mathfrak{A}$ and every $m \geq 1$ there is a structure $\mathfrak{B}$ such that $\mathfrak{A} \cong_m \mathfrak{B}$ and $\mathfrak{B}$ contains at most $m \cdot 2^r$ elements. (*Hint:* Consider the 2^r subsets of A of the form $A_0 \cap \cdots \cap A_{r-1}$, where $A_i = P_i^A$ or $A_i = A - P_i^A$. Choose $\mathfrak{B}$ to be a structure whose corresponding sets have the same number of elements if this number is $<m$, and otherwise have m elements.)

3.9 Exercise. Again let $S = \{P_0, \ldots, P_{r-1}\}$ with unary relation symbols P_i, let $m \geq 1$, and $\varphi \in L_0^S$ be a sentence of quantifier rank $\leq m$. Show:

(a) If φ is satisfiable, then φ is already satisfiable over a domain with at most $m \cdot 2^r$ elements.
(b) $\{\psi \mid \psi \in L^S, \psi \text{ valid}\}$ is R-decidable.

§4. Ehrenfeucht Games

The algebraic description of elementary equivalence is well-suited for many purposes. However, it lacks the intuitive appeal of a game-theoretical characterization due to Ehrenfeucht, which we describe in the present section.

Let S be an arbitrary symbol set and let $\mathfrak{A}$ and $\mathfrak{B}$ be S-structures. To simplify the following formulation we assume $A \cap B = \emptyset$. The *Ehrenfeucht game* $G(\mathfrak{A}, \mathfrak{B})$ corresponding to $\mathfrak{A}$ and $\mathfrak{B}$ is played by two players, I and II, according to the following rules:

Each play of the game begins with player I choosing a natural number $r \geq 1$; r is the number of subsequent moves each player has to make in the course of the play. These subsequent moves are begun by player I, and both players move alternately. Each move consists of choosing an element from $A \cup B$. If player I chooses an element $a_i \in A$ in his ith move, then player II must choose $b_i \in B$ in his ith move. If player I chooses an element $b_i \in B$ in his ith move, then player II must choose $a_i \in A$. After the rth move of player II the play is completed. Altogether some number $r \geq 1$, elements $a_1, \ldots, a_r \in A$ and $b_1, \ldots., b_r \in B$ have been chosen. Player II has won the play iff by $p(a_i) = b_i$ for $i = 1, \ldots, r$ a partial isomorphism from $\mathfrak{A}$ to $\mathfrak{B}$ is defined.

We say that player II has a winning strategy in $G(\mathfrak{A}, \mathfrak{B})$ and write "II *wins* $G(\mathfrak{A}, \mathfrak{B})$" if it is possible for him to win each play. (We omit an exact definition of the notion of winning strategy.)

4.1 Lemma. $\mathfrak{A} \cong_f \mathfrak{B}$ *iff* II *wins* G($\mathfrak{A}$, $\mathfrak{B}$).

This lemma, together with Fraissé's theorem, yields the desired game-theoretical characterization of elementary equivalence:

4.2 Ehrenfeucht's Theorem. *For finite S and arbitrary* $\mathfrak{A}$ *and* $\mathfrak{B}$:

$$\mathfrak{A} \equiv \mathfrak{B} \quad \textit{iff} \text{ II } \textit{wins} \text{ G}(\mathfrak{A}, \mathfrak{B}).$$

PROOF OF 4.1. Suppose $(I_n)_{n\in\mathbb{N}}: \mathfrak{A} \cong_f \mathfrak{B}$. Then $(I'_n)_{n\in\mathbb{N}}: \mathfrak{A} \cong_f \mathfrak{B}$ also, where

$$I'_n := \{p \mid \text{there is } q \in I_n \text{ such that } p \subset q\}.$$

We describe a winning strategy for player II:

If player I chooses the number r at the beginning of a G($\mathfrak{A}$, $\mathfrak{B}$)-play, then for $i = 1, \ldots, r$ player II should choose the elements a_i (or resp. b_i) so that by $p_i(a_j) = b_j$ for $1 \leq j \leq i$ one obtains a partial isomorphism p_i in I'_{r-i}; this is always possible because of the extension properties of partial isomorphisms in $(I'_n)_{n\in\mathbb{N}}$. For $i = r$ it follows that player II wins the play.

Conversely, suppose that player II has a winning strategy in G($\mathfrak{A}$, $\mathfrak{B}$). We define a sequence $(I_n)_{n\in\mathbb{N}}$ as follows:

For $n \in \mathbb{N}$ let

$p \in I_n$ iff $p \in \text{Part}(\mathfrak{A}, \mathfrak{B})$ and there are $j \in \mathbb{N}$ and $a_1, \ldots, a_j \in A$ such that

(i) $\text{dom}(p) = \{a_1, \ldots, a_j\}$;

(ii) there is an $m \geq n$ and a G($\mathfrak{A}$, $\mathfrak{B}$)-play which II plays according to his winning strategy, which player I opens by choosing the number $m + j$, and where in the first j moves the elements $a_1, \ldots, a_j \in A$ and $p(a_1), \ldots, p(a_j) \in B$ are chosen.

From the rules of the game we immediately obtain that $(I_n)_{n\in\mathbb{N}}: \mathfrak{A} \cong_f \mathfrak{B}$. □

CHAPTER XII

Characterizing First-Order Logic

In this final chapter we present some results, due to Lindström [21], which we have already mentioned several times. They show that first-order logic occupies a unique place among logical systems. Indeed we shall prove:

(a) There is no logical system with more expressive power than first-order logic, for which both the compactness theorem and the Löwenheim–Skolem theorem hold (§3).
(b) There is no logical system with more expressive power than first-order logic, for which the Löwenheim–Skolem theorem holds and for which the set of valid sentences is enumerable (§4).

§1. Logical Systems

In the definition of a logical system which follows, we collect several properties which are shared by the logics we have considered so far. As we are mainly interested in semantic aspects we shall speak of a logical system as soon as we have the following: We are given, for every symbol set S, an "abstract" set whose elements play the rôle of S-sentences, and in addition, a relationship between structures and such sentences which corresponds to the satisfaction relation, and determines whether an "abstract" sentence holds in a structure.

1.1 Definition. A *logical system* $\mathscr{L}$ consists of a function L and a binary relation $\models_{\mathscr{L}}$. L associates with every symbol set S a set $L(S)$, the *set of S-sentences* of $\mathscr{L}$. The following properties are required:

(a) If $S_0 \subset S_1$ then $L(S_0) \subset L(S_1)$.
(b) If $\mathfrak{A} \models_{\mathscr{L}} \varphi$ (i.e., if $\mathfrak{A}$ and φ are related under $\models_{\mathscr{L}}$), then, for some S, $\mathfrak{A}$ is an S-structure and $\varphi \in L(S)$.

(c) (*Isomorphism property*) If $\mathfrak{A} \models_{\mathscr{L}} \varphi$ and $\mathfrak{A} \cong \mathfrak{B}$ then $\mathfrak{B} \models_{\mathscr{L}} \varphi$.
(d) (*Reduct property*) If $S_0 \subset S_1$, $\varphi \in L(S_0)$, and $\mathfrak{A}$ is an S_1-structure then

$$\mathfrak{A} \models_{\mathscr{L}} \varphi \quad \text{iff } \mathfrak{A} \upharpoonright S_0 \models_{\mathscr{L}} \varphi.$$

$\mathscr{L}_{\mathrm{I}}$, $\mathscr{L}_{\mathrm{II}}$, $\mathscr{L}_{\mathrm{II}}^{\mathrm{w}}$, $\mathscr{L}_{\omega_1\omega}$, and $\mathscr{L}_Q$ are logical systems. For instance, in the case of $\mathscr{L}_{\mathrm{I}}$ we choose L_{I} to be the function which assigns to a symbol set S the set $L_{\mathrm{I}}(S) := L_0^S$ of first-order S-sentences, and we take $\models_{\mathscr{L}_{\mathrm{I}}}$ to be the usual satisfaction relation between structures and first-order sentences.

If $\mathscr{L}$ is a logical system and $\varphi \in L(S)$, let

$$\mathrm{Mod}_{\mathscr{L}}^S(\varphi) := \{\mathfrak{A} \mid \mathfrak{A} \text{ is an } S\text{-structure and } \mathfrak{A} \models_{\mathscr{L}} \varphi\}.$$

In case S is clear from the context we just write $\mathrm{Mod}_{\mathscr{L}}(\varphi)$.

$\mathrm{Mod}_{\mathscr{L}}^S(\varphi)$ can be regarded as a mathematically precise counterpart to the *meaning* of φ. It suggests the following definition of when a logical system $\mathscr{L}'$ has more expressive power than $\mathscr{L}$, namely, if for every $\mathscr{L}$-sentence φ there is an $\mathscr{L}'$-sentence ψ with the same meaning.

1.2 Definition. Let $\mathscr{L}$ and $\mathscr{L}'$ be logical systems.

(a) *$\mathscr{L}'$ is at least as strong as $\mathscr{L}$* (written: $\mathscr{L} \leq \mathscr{L}'$) iff for every S and every $\varphi \in L(S)$ there is a $\psi \in L'(S)$ such that

$$\mathrm{Mod}_{\mathscr{L}}^S(\varphi) = \mathrm{Mod}_{\mathscr{L}'}^S(\psi).$$

(b) *$\mathscr{L}$ and $\mathscr{L}'$ are equally strong* (written: $\mathscr{L} \sim \mathscr{L}'$) iff $\mathscr{L} \leq \mathscr{L}'$ and $\mathscr{L}' \leq \mathscr{L}$.

EXAMPLES. $\mathscr{L}_{\mathrm{I}} \leq \mathscr{L}_{\mathrm{II}}^{\mathrm{w}}$; $\mathscr{L}_{\mathrm{II}}^{\mathrm{w}} \leq \mathscr{L}_{\mathrm{II}}$; not $\mathscr{L}_{\mathrm{II}} \leq \mathscr{L}_{\mathrm{II}}^{\mathrm{w}}$ (cf. IX.1.7); $\mathscr{L}_{\mathrm{II}}^{\mathrm{w}} \leq \mathscr{L}_{\omega_1\omega}$ (cf. IX.2.7); not $\mathscr{L}_{\mathrm{II}}^{\mathrm{w}} \leq \mathscr{L}_Q$ (proof!) and not $\mathscr{L}_Q \leq \mathscr{L}_{\mathrm{II}}^{\mathrm{w}}$.

On our abstract level we now formulate some properties of logical systems which are known to hold for the systems we have considered so far.

Boole($\mathscr{L}$) ("$\mathscr{L}$ is closed under propositional ("Boolean") connectives") if (1) and (2) are satisfied.

(1) Given S and $\varphi \in L(S)$, there is a $\chi \in L(S)$ such that for every S-structure $\mathfrak{A}$:

$$\mathfrak{A} \models_{\mathscr{L}} \chi \quad \text{iff not } \mathfrak{A} \models_{\mathscr{L}} \varphi.$$

(2) Given S and $\varphi, \psi \in L(S)$, there is a $\chi \in L(S)$ such that for every S-structure $\mathfrak{A}$:

$$\mathfrak{A} \models_{\mathscr{L}} \chi \quad \text{iff} \qquad \mathfrak{A} \models_{\mathscr{L}} \varphi \text{ or } \mathfrak{A} \models_{\mathscr{L}} \psi.$$

If Boole($\mathscr{L}$) holds then let $\neg \varphi$ and $(\varphi \vee \psi)$ stand for formulas χ in the sense of (1) and (2) above. $(\varphi \wedge \psi)$, $(\varphi \rightarrow \psi)$, ... are used analogously.

Rel($\mathscr{L}$) ("$\mathscr{L}$ permits relativization", cf. VIII.2):

For S, $\varphi \in L(S)$, and unary U there is a $\psi \in L(S \cup \{U\})$ such that

$$(\mathfrak{A}, U^A) \models_{\mathscr{L}} \psi \quad \text{iff } [U^A]^{\mathfrak{A}} \models_{\mathscr{L}} \varphi$$

for all S-structures $\mathfrak{A}$ and all S-closed subsets U^A of A. ($[U^A]^{\mathfrak{A}}$ is the substructure of $\mathfrak{A}$ with domain U^A.)

If Rel($\mathscr{L}$) holds let φ^U be a formula ψ with the above property.

Elim($\mathscr{L}$) ("$\mathscr{L}$ allows elimination of function symbols and constants"):

If S is a symbol set and S^r is chosen as in VIII.1, then for any $\varphi \in L(S)$ there is a $\psi \in L(S^r)$ such that for all S-structures $\mathfrak{A}$:

$$\mathfrak{A} \models_{\mathscr{L}} \varphi \quad \text{iff } \mathfrak{A}^r \models_{\mathscr{L}} \psi.$$

(For the definition of $\mathfrak{A}^r$ cf. also VIII.1). If Elim($\mathscr{L}$) then we write φ^r for a formula ψ with the above property.

1.3 Definition. A logical system $\mathscr{L}$ is said to be *regular* if it satisfies the properties Boole($\mathscr{L}$), Rel($\mathscr{L}$), and Elim($\mathscr{L}$).

All logical systems which we have hitherto considered are regular. In the case of $\mathscr{L}_{\mathrm{I}}$ we verified Elim($\mathscr{L}_{\mathrm{I}}$) and Rel($\mathscr{L}_{\mathrm{I}}$) in VIII.1 and VIII.2. The arguments given there can also be applied without difficulty for the other logical systems.

We tacitly adopt some semantic notions whose definition can be extended from $\mathscr{L}_1$ to the general case in a straightforward manner. For example, $\varphi \in L(S)$ is said to be *satisfiable* if $\mathrm{Mod}^S_{\mathscr{L}}(\varphi) \neq \emptyset$, and *valid* if $\mathrm{Mod}^S_{\mathscr{L}}(\varphi)$ is the class of all S-structures. If $\Phi \subset L(S)$ then $\Phi \models_{\mathscr{L}} \varphi$ means that every model of Φ (in the sense of $\models_{\mathscr{L}}$) is a model of φ. Note that these definitions refer to a fixed symbol set S. However, using the reduct property 1.1(d) one can argue as in III.5.3 to show that they do not depend on S. In the sequel similar applications of 1.1(d) will be made without explicit mention.

We introduce the following abbreviations:

LöSko($\mathscr{L}$) ("The Löwenheim–Skolem theorem holds for $\mathscr{L}$"):
Every satisfiable sentence of $\mathscr{L}$ has an at most countable model.

Comp($\mathscr{L}$) ("The compactness theorem holds for $\mathscr{L}$"):
If Φ is a set of sentences of $\mathscr{L}$ such that every finite subset of Φ is satisfiable, then Φ is satisfiable.

In this terminology the result of Lindström mentioned in (a) reads as follows:

Let $\mathscr{L}$ be a regular logical system such that $\mathscr{L}_{\mathrm{I}} \leq \mathscr{L}$, LöSko($\mathscr{L}$), and Comp($\mathscr{L}$). Then $\mathscr{L} \sim \mathscr{L}_{\mathrm{I}}$.

Before embarking on the proof (in §3) we derive in the next section some properties of logical systems for which the compactness theorem holds.

1.4 Exercise. Let $\mathscr{L}$ be given by

(i) $L(S) := \{\varphi \mid \varphi$ is an L^S_{II}-sentence of the form $\exists X_0 \ldots \exists X_{n-1}\psi$, where ψ does not contain a second-order quantifier$\}$.

(ii) For $\varphi \in L(S)$ and any S-structure $\mathfrak{A}$,

$$\mathfrak{A} \models_{\mathscr{L}} \varphi \quad \text{iff } \mathfrak{A} \models_{\mathscr{L}_{\text{II}}} \varphi.$$

Show:

(a) $\mathscr{L}$ is a logical system.
(b) LöSko($\mathscr{L}$), Comp($\mathscr{L}$), Rel($\mathscr{L}$), and Elim($\mathscr{L}$).
(c) Not Boole($\mathscr{L}$).
(d) $\mathscr{L}_{\text{I}} \leq \mathscr{L}$ but not $\mathscr{L} \leq \mathscr{L}_{\text{I}}$.

§2. Compact Regular Logical Systems

In this section $\mathscr{L}$ will always be a *regular logical system such that* $\mathscr{L}_{\text{I}} \leq \mathscr{L}$.

For a first-order S-sentence φ, let φ^* be a sentence in $L(S)$ which has the same models as φ, i.e.,

$$\text{Mod}^S_{\mathscr{L}_{\text{I}}}(\varphi) = \text{Mod}^S_{\mathscr{L}}(\varphi^*).$$

For a set Φ of first-order S-sentences define $\Phi^* := \{\varphi^* \mid \varphi \in \Phi\}$.

If $\mathscr{L}$ is compact, i.e., Comp($\mathscr{L}$) holds, then also the compactness theorem for the consequence relation holds, as can be shown similarly to the first-order case:

2.1 Lemma. *Suppose* Comp($\mathscr{L}$), *and let* $\Phi \cup \{\varphi\} \subset L(S)$ *and* $\Phi \models_{\mathscr{L}} \varphi$. *Then there is a finite subset* Φ_0 *of* Φ *such that* $\Phi_0 \models_{\mathscr{L}} \varphi$.

Proof. Choose $\neg\varphi$ by Boole($\mathscr{L}$). Then $\Phi \cup \{\neg\varphi\}$ is not satisfiable. By Comp($\mathscr{L}$) there is a finite subset Φ_0 of Φ so that $\Phi_0 \cup \{\neg\varphi\}$ is not satisfiable, i.e., $\Phi_0 \models_{\mathscr{L}} \varphi$. □

As a further property of compact logics, we show that the meaning of an $L(S)$-sentence only depends on finitely many symbols from S:

2.2 Lemma. *Suppose* Comp($\mathscr{L}$) *and* $\psi \in L(S)$. *Then there is a finite subset* S_0 *of* S *such that for all* S*-structures* $\mathfrak{A}$ *and* $\mathfrak{B}$,

$$\textit{if} \quad \mathfrak{A} \upharpoonright S_0 \cong \mathfrak{B} \upharpoonright S_0 \quad \textit{then} \quad (\mathfrak{A} \models_{\mathscr{L}} \psi \quad \textit{iff} \quad \mathfrak{B} \models_{\mathscr{L}} \psi).$$

Proof. We restrict ourselves to the case where S is relational (the case we shall subsequently need). There is no difficulty in extending the proof to arbitrary symbol sets.

Choose new unary symbols U, V, and f. Let Φ consist of the following first-order $S \cup \{U, V, f\}$-sentences, which say that f is an isomorphism between the substructure induced on U and the substructure induced on V:

$$\exists x Ux, \exists x Vx,$$

$$\forall x(Ux \rightarrow Vfx), \forall y(Vy \rightarrow \exists x(Ux \wedge fx \equiv y)),$$

$$\forall x \forall y((Ux \wedge Uy \wedge fx \equiv fy) \rightarrow x \equiv y),$$

and for every $R \in S$, R n-ary,

$$\forall x_0 \ldots \forall x_{n-1}((Ux_0 \wedge \cdots \wedge Ux_{n-1}) \rightarrow (Rx_0 \ldots x_{n-1} \leftrightarrow Rfx_0 \ldots fx_{n-1})).$$

Then, firstly,

$$\Phi^* \models_{\mathscr{L}} \psi^U \leftrightarrow \psi^V. \tag{1}$$

In fact, if $\mathfrak{A}$ is an S-structure and $(\mathfrak{A}, U^A, V^A, f^A) \models_{\mathscr{L}} \Phi^*$, i.e., $(\mathfrak{A}, U^A, V^A, f^A) \models \Phi$, then U^A and V^A are nonempty and $f^A \upharpoonright U^A$ is an isomorphism from $[U^A]^{\mathfrak{A}}$ to $[V^A]^{\mathfrak{A}}$. By the isomorphism property (cf. 1.1(c)) we have

$$[U^A]^{\mathfrak{A}} \models_{\mathscr{L}} \psi \quad \text{iff } [V^A]^{\mathfrak{A}} \models_{\mathscr{L}} \psi,$$

that is, by Rel($\mathscr{L}$),

$$(\mathfrak{A}, U^A) \models_{\mathscr{L}} \psi^U \quad \text{iff } (\mathfrak{A}, V^A) \models_{\mathscr{L}} \psi^V.$$

Using the reduct property and Boole($\mathscr{L}$) we obtain

$$(\mathfrak{A}, U^A, V^A, f^A) \models_{\mathscr{L}} \psi^U \leftrightarrow \psi^V.$$

Thus (1) is proved. By Comp($\mathscr{L}$) there is a finite subset Φ_0 of Φ such that

$$\Phi_0^* \models_{\mathscr{L}} \psi^U \leftrightarrow \psi^V. \tag{2}$$

Since Φ_0 consists of first-order sentences, we may choose a finite subset S_0 of S such that Φ_0 consists of S_0-sentences. We show that S_0 has the desired properties. Suppose $\mathfrak{A}$ and $\mathfrak{B}$ are S-structures and π: $\mathfrak{A} \upharpoonright S_0 \cong \mathfrak{B} \upharpoonright S_0$, where we assume $A \cap B = \varnothing$. (Otherwise we can take an isomorphic copy of $\mathfrak{B}$ and use the isomorphism property.) We define over $C := A \cup B$ an $S \cup \{U, V, f\}$-structure $(\mathfrak{C}, U^C, V^C, f^C)$ as follows (note that S is relational):

$$R^C := R^A \cup R^B \quad \text{for } R \in S,$$

$$U^C := A,$$

$$V^C := B,$$

$$f^C \text{ such that } f^C \upharpoonright U^C = \pi.$$

Then $(\mathfrak{C}, U^C, V^C, f^C)$ is a model of Φ_0, i.e. $(\mathfrak{C}, U^C, V^C, f^C) \models_{\mathscr{L}} \Phi_0^*$. Hence by (2),

$$(\mathfrak{C}, U^C, V^C, f^C) \models_{\mathscr{L}} \psi^U \leftrightarrow \psi^V,$$

and therefore, using $[U^C]^{\mathfrak{C}} = \mathfrak{A}$ and $[V^C]^{\mathfrak{C}} = \mathfrak{B}$,

$$\mathfrak{A} \models_{\mathscr{L}} \psi \quad \text{iff } \mathfrak{B} \models_{\mathscr{L}} \psi.$$

□

Two S-structures $\mathfrak{A}$ and $\mathfrak{B}$ are said to be *$\mathscr{L}$-equivalent* (written: $\mathfrak{A} \equiv_{\mathscr{L}} \mathfrak{B}$), if for all $\psi \in L(S)$,

$$\mathfrak{A} \models_{\mathscr{L}} \psi \quad \text{iff } \mathfrak{B} \models_{\mathscr{L}} \psi.$$

For $\mathfrak{A} \equiv_{\mathscr{L}_{\mathrm{I}}} \mathfrak{B}$ we continue to write simply $\mathfrak{A} \equiv \mathfrak{B}$. Clearly, if $\mathscr{L} \sim \mathscr{L}_{\mathrm{I}}$ then $\mathfrak{A} \equiv \mathfrak{B}$ implies $\mathfrak{A} \equiv_{\mathscr{L}} \mathfrak{B}$. We show that the converse holds for compact $\mathscr{L}$.

2.3 Lemma. *Assume* Comp($\mathscr{L}$) *and suppose that* $\mathfrak{A} \equiv \mathfrak{B}$ *implies* $\mathfrak{A} \equiv_{\mathscr{L}} \mathfrak{B}$ *for arbitrary* $\mathfrak{A}$, $\mathfrak{B}$. *Then* $\mathscr{L} \sim \mathscr{L}_{\mathrm{I}}$.

Proof. Since $\mathscr{L}_{\mathrm{I}} \leq \mathscr{L}$, given S and $\psi \in L(S)$, we must show that there is a first-order S-sentence φ such that

$$\mathrm{Mod}_{\mathscr{L}_{\mathrm{I}}}(\varphi) = \mathrm{Mod}_{\mathscr{L}}(\psi).$$

Suppose ψ is satisfiable. (Otherwise we let $\varphi := \forall x \neg x \equiv x$.) First, we claim

$$\text{(1)}\quad \text{For every } \mathfrak{A} \in \mathrm{Mod}_{\mathscr{L}}(\psi) \text{ there is } \varphi_{\mathfrak{A}} \in L_0^S \text{ such that} \quad \mathfrak{A} \models \varphi_{\mathfrak{A}} \quad \text{and} \quad \varphi_{\mathfrak{A}}^* \models_{\mathscr{L}} \psi.$$

To show this we let $\mathfrak{A} \in \mathrm{Mod}_{\mathscr{L}}(\psi)$. Then for the first-order theory $\mathrm{Th}(\mathfrak{A})$ of $\mathfrak{A}$,

$$\mathrm{Th}(\mathfrak{A})^* \models_{\mathscr{L}} \psi.$$

For if $\mathfrak{B} \models_{\mathscr{L}} \mathrm{Th}(\mathfrak{A})^*$, i.e., $\mathfrak{B} \models \mathrm{Th}(\mathfrak{A})$, then $\mathfrak{B} \equiv \mathfrak{A}$ by hypothesis; therefore $\mathfrak{B} \equiv_{\mathscr{L}} \mathfrak{A}$ and hence $\mathfrak{B} \models_{\mathscr{L}} \psi$.

Since Comp($\mathscr{L}$), there are r and $\varphi_0, \ldots, \varphi_r \in \mathrm{Th}(\mathfrak{A})$ such that $\{\varphi_0^*, \ldots, \varphi_r^*\} \models_{\mathscr{L}} \psi$. We set $\varphi_{\mathfrak{A}} := \varphi_0 \wedge \cdots \wedge \varphi_r$. Then $\varphi_{\mathfrak{A}} \in \mathrm{Th}(\mathfrak{A})$, i.e., $\mathfrak{A} \models \varphi_{\mathfrak{A}}$, and $\varphi_{\mathfrak{A}}^* \models_{\mathscr{L}} \psi$. Thus (1) is proved.

From (1) we immediately obtain

$$\text{(2)}\qquad \mathrm{Mod}_{\mathscr{L}}(\psi) = \bigcup_{\mathfrak{A} \in \mathrm{Mod}_{\mathscr{L}}(\psi)} \mathrm{Mod}_{\mathscr{L}}(\varphi_{\mathfrak{A}}^*).$$

Now we show that there are $\mathfrak{A}_0, \ldots, \mathfrak{A}_n \in \mathrm{Mod}_{\mathscr{L}}(\psi)$ such that

$$\text{(3)}\qquad \mathrm{Mod}_{\mathscr{L}}(\psi) = \mathrm{Mod}_{\mathscr{L}}(\varphi_{\mathfrak{A}_0}^*) \cup \cdots \cup \mathrm{Mod}_{\mathscr{L}}(\varphi_{\mathfrak{A}_n}^*).$$

Otherwise, for arbitrary $\mathfrak{A}_0, \ldots, \mathfrak{A}_m$ from $\mathrm{Mod}_{\mathscr{L}}(\psi)$ we would have

$$\mathrm{Mod}_{\mathscr{L}}(\psi) \supsetneqq \mathrm{Mod}_{\mathscr{L}}(\varphi_{\mathfrak{A}_0}^*) \cup \cdots \cup \mathrm{Mod}_{\mathscr{L}}(\varphi_{\mathfrak{A}_m}^*),$$

and thus every finite subset of $\{\psi\} \cup \{\neg \varphi_{\mathfrak{A}}^* \mid \mathfrak{A} \in \mathrm{Mod}_{\mathscr{L}}(\psi)\}$ would be satisfiable. By Comp($\mathscr{L}$) the whole set would be satisfiable, in contradiction to (2).

Writing $\mathrm{Mod}_{\mathscr{L}}(\psi)$ as in (3) we obtain

$$\begin{aligned} \mathrm{Mod}_{\mathscr{L}}(\psi) &= \mathrm{Mod}_{\mathscr{L}_{\mathrm{I}}}(\varphi_{\mathfrak{A}_0}) \cup \cdots \cup \mathrm{Mod}_{\mathscr{L}_{\mathrm{I}}}(\varphi_{\mathfrak{A}_n}) \\ &= \mathrm{Mod}_{\mathscr{L}_{\mathrm{I}}}(\varphi_{\mathfrak{A}_0} \vee \cdots \vee \varphi_{\mathfrak{A}_n}). \end{aligned}$$

Hence for $\varphi := \varphi_{\mathfrak{A}_0} \vee \cdots \vee \varphi_{\mathfrak{A}_n}$ we have $\mathrm{Mod}_{\mathscr{L}_{\mathrm{I}}}(\varphi) = \mathrm{Mod}_{\mathscr{L}}(\psi)$. □

§3. Lindström's First Theorem

We now have all tools available to obtain the following characterization of first-order logic.

3.1 Lindström's First Theorem. *Let $\mathcal{L}$ be a regular logical system such that $\mathcal{L}_{\mathrm{I}} \leq \mathcal{L}$. Then*

$$\text{LöSko}(\mathcal{L}) \text{ and } \text{Comp}(\mathcal{L}) \text{ imply } \mathcal{L} \sim \mathcal{L}_{\mathrm{I}}.$$

PROOF. Given $\mathcal{L}$ satisfying the conditions of the theorem we must show that $\mathcal{L} \leq \mathcal{L}_{\mathrm{I}}$. By 2.3 it is sufficient to prove for all S:

$(+)$ For all S-structures $\mathfrak{A}$ and $\mathfrak{B}$, if $\mathfrak{A} \equiv \mathfrak{B}$ then $\mathfrak{A} \equiv_{\mathcal{L}} \mathfrak{B}$.

We can restrict ourselves here to relational symbol sets. Assuming that $(+)$ has been established for relational symbol sets, we can then give the following argument for arbitrary S:

Suppose $\mathfrak{A} \equiv \mathfrak{B}$. Considering S^r, $\mathfrak{A}^r$, and $\mathfrak{B}^r$ we obtain, first of all, $\mathfrak{A}^r \equiv \mathfrak{B}^r$ (cf. VIII.1.7). Then for the relational symbol set S^r, $(+)$ yields $\mathfrak{A}^r \equiv_{\mathcal{L}} \mathfrak{B}^r$. Finally, using Elim($\mathcal{L}$), we have for arbitrary $\psi \in L(S)$:

$$\begin{aligned} \mathfrak{A} \models_{\mathcal{L}} \psi \quad &\text{iff } \mathfrak{A}^r \models_{\mathcal{L}} \psi^r \\ &\text{iff } \mathfrak{B}^r \models_{\mathcal{L}} \psi^r \quad (\text{since } \mathfrak{A}^r \equiv_{\mathcal{L}} \mathfrak{B}^r) \\ &\text{iff } \mathfrak{B} \models_{\mathcal{L}} \psi, \end{aligned}$$

and thus $\mathfrak{A} \equiv_{\mathcal{L}} \mathfrak{B}$.

For relational S we now prove $(+)$. Assume, for contradiction, that we have S-structures $\mathfrak{A}$ and $\mathfrak{B}$, and $\psi \in L(S)$ such that

$$(1) \qquad \mathfrak{A} \equiv \mathfrak{B}, \qquad \mathfrak{A} \models_{\mathcal{L}} \psi, \qquad \mathfrak{B} \models_{\mathcal{L}} \neg\psi.$$

Corresponding to ψ we choose a finite subset S_0 of S as in lemma 2.2, so that the meaning of ψ only depends on the symbols in S_0.

Since $\mathfrak{A} \upharpoonright S_0 \equiv \mathfrak{B} \upharpoonright S_0$ holds, $\mathfrak{A} \upharpoonright S_0$ and $\mathfrak{B} \upharpoonright S_0$ are finitely isomorphic by Fraissé's theorem, and hence for suitable $(I_n)_{n \in \mathbb{N}}$ we have

$$(2) \qquad (I_n)_{n \in \mathbb{N}}: \mathfrak{A} \upharpoonright S_0 \cong_f \mathfrak{B} \upharpoonright S_0, \qquad \mathfrak{A} \models_{\mathcal{L}} \psi, \qquad \mathfrak{B} \models_{\mathcal{L}} \neg\psi.$$

The central idea of the proof is to apply Comp($\mathcal{L}$) and LöSko($\mathcal{L}$) to obtain structures $\mathfrak{A}'$ and $\mathfrak{B}'$ which are at most countable and for which

$$(3) \qquad \mathfrak{A}' \upharpoonright S_0 \cong_p \mathfrak{B}' \upharpoonright S_0, \qquad \mathfrak{A}' \models_{\mathcal{L}} \psi, \quad \text{and} \quad \mathfrak{B}' \models_{\mathcal{L}} \neg\psi.$$

Once we have (3) we get the desired contradiction: The at most countable, partially isomorphic structures $\mathfrak{A}' \upharpoonright S_0$ and $\mathfrak{B}' \upharpoonright S_0$ are isomorphic (cf. XI.1.5(d)). Hence by the choice of S_0

$$\mathfrak{A}' \models_{\mathcal{L}} \psi \quad \text{iff } \mathfrak{B}' \models_{\mathcal{L}} \psi,$$

and this contradicts $\mathfrak{A}' \models_{\mathcal{L}} \psi$ and $\mathfrak{B}' \models_{\mathcal{L}} \neg\psi$ of (3).

In order to proceed from (2) to (3) we give a suitable "description" of (2) in $\mathcal{L}$. We may assume that $A \cap B = \emptyset$ for $\mathfrak{A}$, $\mathfrak{B}$ in (2) (otherwise take an isomorphic copy of $\mathfrak{B}$). Let the symbol set S^+ be formed from S by adding the following new symbols: a unary function symbol f and relation symbols P, U, V (unary), $<$, I (binary), and G (ternary). We define an S^+-structure $\mathfrak{C}$

in which (2) can be suitably described. In particular $\mathfrak{C}$ includes $\mathfrak{A}$ and $\mathfrak{B}$ and also contains the partial isomorphisms of the I_n in (2): Set

(a) $C = A \cup B \cup \mathbb{N} \cup \bigcup_{n \in \mathbb{N}} I_n$;
(b) $U^C = A$ and $[U^C]^{\mathfrak{C} \upharpoonright S} = \mathfrak{A}$;
(c) $V^C = B$ and $[V^C]^{\mathfrak{C} \upharpoonright S} = \mathfrak{B}$;

((b) and (c) are possible since $A \cap B = \emptyset$ and since S is relational.)

(d) $<^C$ the natural ordering relation on $\mathbb{N}$ and $f^C \upharpoonright \mathbb{N}$ the predecessor function on $\mathbb{N}$, i.e., $f^C(n + 1) = n$ and $f^C(0) = 0$;
(e) $P^C = \bigcup_{n \in \mathbb{N}} I_n$;
(f) $I^C np$ iff $n \in \mathbb{N}$ and $p \in I_n$;
(g) $G^C pab$ iff $P^C p$, $a \in \mathrm{dom}(p)$ and $p(a) = b$.

This is illustrated in Fig. 12.1

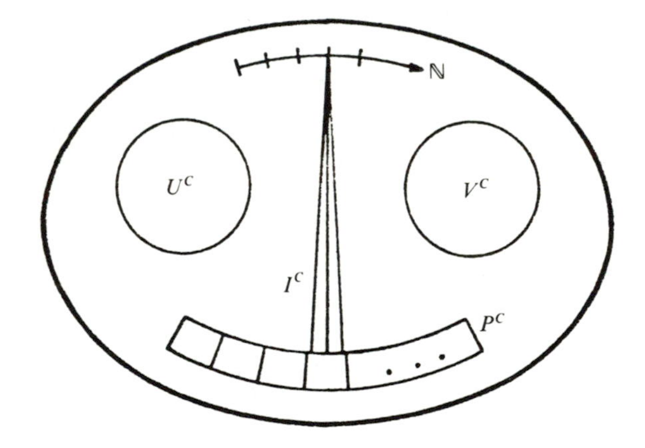

Figure 12.1

$\mathfrak{C}$ is then a model of the conjunction χ of the following finite set of sentences of $L(S^+)$. (Since $\mathscr{L}_{\mathrm{I}} \leq \mathscr{L}$ we use first-order sentences as an intuitive notation for the corresponding sentences of $\mathscr{L}$.)

(i) $\forall p(Pp \rightarrow \forall x \, \forall y(Gpxy \rightarrow (Ux \wedge Vy)))$.
(ii) $\forall p(Pp \rightarrow \forall x \, \forall x' \, \forall y \, \forall y'((Gpxy \wedge Gpx'y') \rightarrow (x \equiv x' \leftrightarrow y \equiv y')))$.
(iii) For every $R \in S_0$, R n-ary:

$$\forall p(Pp \rightarrow \forall x_0 \ldots \forall x_{n-1} \, \forall y_0 \ldots \forall y_{n-1}((Gpx_0 y_0 \wedge \cdots \wedge Gpx_{n-1} y_{n-1}) \rightarrow (Rx_0 \ldots x_{n-1} \leftrightarrow Ry_0 \ldots y_{n-1}))).$$

(In effect (i), (ii), and (iii) say that for a fixed $p \in P$, $Gp \cdot\cdot$ describes the graph of a partial isomorphism of the S_0-substructure induced on U to the S_0-substructure induced on V.)

(iv) The axioms of Φ_{pord} (for partially defined orderings) and

$$\forall x(\exists y\, y < x \rightarrow (fx < x \wedge \neg \exists z(fx < z \wedge z < x)))$$

(i.e., (field $<$, $<$) is an ordering with predecessor function f).

(v) $\forall x(\exists y(x < y \vee y < x) \rightarrow \exists p(Pp \wedge Ixp))$ (i.e., if x is in the field of $<$ then $I_x = \{p \mid Pp \wedge Ixp\}$ is not empty).

(vi) $\forall x\, \forall p\, \forall u((fx < x \wedge Ixp \wedge Uu)$
$\rightarrow \exists q\, \exists v(Ifxq \wedge Gquv \wedge \forall x'\, \forall y'(Gpx'y' \rightarrow Gqx'y')))$
(the forth-property).

(vii) An analogous sentence for the back-property.

(viii) $\exists x\, Ux \wedge \exists y\, Vy \wedge \psi^U \wedge (\neg\psi)^V$ (note that $U^C = A$, $V^C = B$, $\mathfrak{A} \models_{\mathscr{L}} \psi$, $\mathfrak{B} \models_{\mathscr{L}} \neg\psi$).

We choose a new constant c and write $f^0c, f^1c, f^2c, \ldots$ for the terms $c, fc, ffc, \ldots$. Then we let

$$\Psi = \{\chi\} \cup \{\ldots ffc < fc < c\},$$

i.e.,

$$\Psi = \{\chi\} \cup \{f^{n+1}c < f^nc \mid n \in \mathbb{N}\}.$$

Every finite subset of Ψ has a model, namely $\mathfrak{C}' = (\mathfrak{C}, c^{C'})$, where $c^{C'}$ is a sufficiently large natural number. By Comp($\mathscr{L}$) there is a model of Ψ, say $(\mathfrak{D}, c^D)$.

D contains an infinite descending chain, namely $\ldots (f^2c)^D <^D (fc)^D <^D c^D$. For what follows we need a *countable* model with this property. LöSko($\mathscr{L}$) does not directly help us here because it only applies to single sentences, whereas Ψ is an infinite set of sentences. We circumvent this difficulty using a new unary relation symbol Q: Let ϑ be the $L(S^+ \cup \{c, Q\})$-sentence

$$\vartheta = Qc \wedge \forall x(Qx \rightarrow (fx < x \wedge Qfx))$$

("Q contains c and every element of Q possesses an immediate $<$-predecessor which also belongs to Q"; thus Q is a subset of the field of $<$).

If $Q^D := \{(f^nc)^D \mid n \in \mathbb{N}\}$ then

$$(\mathfrak{D}, c^D, Q^D) \models_{\mathscr{L}} \chi \wedge \vartheta,$$

i.e., $\chi \wedge \vartheta$ is satisfiable. Therefore, by LöSko($\mathscr{L}$), there is an at most countable model $(\mathfrak{E}, c^E, Q^E)$ of $\chi \wedge \vartheta$. Since (viii) holds in $\mathfrak{E}$, $U^E \neq \emptyset$, $V^E \neq \emptyset$, and because S is relational, U^E and V^E are domains of substructures. We set

$$\mathfrak{A}' := [U^E]^{\mathfrak{E} \upharpoonright S}, \qquad \mathfrak{B}' := [V^E]^{\mathfrak{E} \upharpoonright S},$$

and show that the at most countable structures $\mathfrak{A}'$ and $\mathfrak{B}'$ satisfy the conditions in (3). By (viii), $\mathfrak{E} \models_{\mathscr{L}} \psi^U$ and $\mathfrak{E} \models_{\mathscr{L}} (\neg\psi)^V$, and from this we obtain

(4) $$\mathfrak{A}' \models_{\mathscr{L}} \psi, \qquad \mathfrak{B}' \models_{\mathscr{L}} \neg\psi.$$

From (i), (ii), and (iii) we know that to every $p \in P^E$ there corresponds a partial isomorphism from $\mathfrak{A}' \upharpoonright S_0$ to $\mathfrak{B}' \upharpoonright S_0$, which we also denote by p.

Since $(\mathfrak{E}, c^E, Q^E) \models \vartheta$, for every $n \in \mathbb{N}$ the element $e_n := (f^n c)^E$ belongs to Q^E and the e_n form a descending chain $\ldots e_3 <^E e_2 <^E e_1 <^E e_0$. Let

$$I := \{p \mid \text{there is an } n \text{ such that } I^E e_n p\}.$$

Using (v) we see that $I \neq \emptyset$, and using (vi) and (vii), that I has the back- and forth-property. For example, to verify the forth-property one can reason as follows: if $p \in I$, say $I^E e_n p$, and $a \in A' = U^E$, then by (vi), there is q such that $I^E e_{n+1} q$ (thus $q \in I$), $q \supset p$, and $a \in \mathrm{dom}(q)$. To summarize, we have

(5) $$I: \mathfrak{A}' \upharpoonright S_0 \cong_p \mathfrak{B}' \upharpoonright S_0.$$

(4) and (5) yield (3) and thus 3.1 is proved. □

By closer inspection of the above proof we can obtain the following lemma, which we shall need in the next section.

3.2 Lemma. *Let $\mathscr{L}$ be a regular logical system such that $\mathscr{L}_\mathrm{I} \leq \mathscr{L}$ and* LöSko($\mathscr{L}$). *Assume that for some finite set S of relation symbols there is $\psi \in L(S)$ such that for every $m \in \mathbb{N}$ there are S-structures $\mathfrak{A}_m$ and $\mathfrak{B}_m$ with*

(*) $$\mathfrak{A}_m \cong_m \mathfrak{B}_m, \qquad \mathfrak{A}_m \models_{\mathscr{L}} \psi \quad \text{and} \quad \mathfrak{B}_m \models_{\mathscr{L}} \neg\psi.$$

Then in $\mathscr{L}$, the class of finite orderings can be described in the following sense: There is a finite S_1 containing symbols $<$, c and a sentence $\chi_1 \in L(S_1)$ such that (a) *and* (b) *hold:*

(a) *If $\mathfrak{A} \models_{\mathscr{L}} \chi_1$ then $(A, <^A)$ is a partially defined ordering and c^A is an element of the field with only finitely many $<^A$-predecessors.*
(b) *For every $m \in \mathbb{N}$ there is a model $\mathfrak{A}$ of χ_1 in which c^A has at least m $<^A$-predecessors.*

Proof. Let S, ψ be given as above. Choose χ and c as in the preceding proof (taking now $S_0 = S$) and set $\chi_1 = \chi \wedge$ "c is in the field of $<$".

We first prove (b). For a given m, let $\mathfrak{A}_m$ and $\mathfrak{B}_m$ be as in (*) and let $(I_n)_{n \leq m}: \mathfrak{A}_m \cong \mathfrak{B}_m$. Define $\mathfrak{C}$ as in the proof of 3.1 except for the following obvious modifications:

(i) $<^C$ is the natural ordering on $\{0, \ldots, m\}$;
(ii) $P^C = \bigcup_{n \leq m} I_n$.

For $c^C := m$ the structure $(\mathfrak{C}, c^C)$ is a model of χ_1, and c^C has m $<^C$-predecessors.

Proof of (a): Suppose there is a model $(\mathfrak{D}, c^D)$ of χ_1 in which c^D has infinitely many $<^D$-predecessors. From $(\mathfrak{D}, c^D)$ we can proceed as in the proof of 3.1 to obtain isomorphic structures $\mathfrak{A}'$ and $\mathfrak{B}'$ such that $\mathfrak{A}' \models \psi$ and $\mathfrak{B}' \models \neg\psi$, a contradiction. □

Lindström's theorem 3.1 characterizes first-order logic in the following sense: Among the regular logical systems there is none of greater expressive power which still satisfies the compactness theorem and the Löwenheim–Skolem theorem.

If one considers the defining properties of regular logical systems $\mathscr{L}$, the properties Rel($\mathscr{L}$) and Elim($\mathscr{L}$) do not seem as fundamental as the others. An analysis of the proof of 3.1 shows that both these properties were used to speak about two structures $\mathfrak{A}$ and $\mathfrak{B}$ by placing them together in the structure $\mathfrak{C}$. There are alternative properties that can be used for the same purpose (cf. [2]). But if there is no substitute at all for Rel($\mathscr{L}$) and Elim($\mathscr{L}$), then there are counterexamples to 3.1.

§4. Lindström's Second Theorem

In our considerations of logical systems we now pay special attention to syntactic aspects. In this connection we recall the following properties of first-order logic: For a decidable symbol set S

the S-sentences are concrete finite symbol strings and the set of S-sentences is decidable;

operations such as negation, relativization, and the elimination of function symbols can be carried out effectively;

there exists an adequate proof calculus, and therefore the set of valid S-sentences is enumerable.

We shall consider these aspects for logical systems in general, thereby arriving at the concept of an effective logical system. Within this framework we can then formulate and prove the result of Lindström mentioned in the introduction to this chapter under (b).

When speaking of a decidable set we understand it to be a set of words over a suitable alphabet which is R-decidable in the sense of X.2.5.

4.1 Definition. Let $\mathscr{L}$ be a logical system. $\mathscr{L}$ is called an *effective logical system* if for every decidable symbol set S the set $L(S)$ is decidable, and for every $\varphi \in L(S)$ there is a finite subset S_0 of S such that $\varphi \in L(S_0)$.

4.2 Definition. Let $\mathscr{L}$ and $\mathscr{L}'$ be effective logical systems.

(a) $\mathscr{L} \leq_{\text{eff}} \mathscr{L}'$ iff for every decidable S there is a computable function $*$ which associates with every $\varphi \in L(S)$ a sentence $\varphi^* \in L'(S)$ such that $\text{Mod}^S_{\mathscr{L}}(\varphi) = \text{Mod}^S_{\mathscr{L}'}(\varphi^*)$.

(b) $\mathscr{L} \sim_{\text{eff}} \mathscr{L}'$ iff ($\mathscr{L} \leq_{\text{eff}} \mathscr{L}'$ and $\mathscr{L}' \leq_{\text{eff}} \mathscr{L}$).

$\mathscr{L}_{\text{I}}$, $\mathscr{L}^{\text{w}}_{\text{II}}$, $\mathscr{L}_{\text{II}}$, and $\mathscr{L}_Q$ are effective logical systems, but $\mathscr{L}_{\omega_1\omega}$ is not. We have, for instance, $\mathscr{L}_{\text{I}} \leq_{\text{eff}} \mathscr{L}^{\text{w}}_{\text{II}}$, $\mathscr{L}^{\text{w}}_{\text{II}} \leq_{\text{eff}} \mathscr{L}_{\text{II}}$.

4.3 Definition. Let $\mathscr{L}$ be a logical system. $\mathscr{L}$ is said to be *effectively regular* if $\mathscr{L}$ is effective and if the following effective analogues of Boole($\mathscr{L}$), Rel($\mathscr{L}$), and Elim($\mathscr{L}$) hold.

For every decidable symbol set S:

(i) There is a computable function which assigns to every $\varphi \in L(S)$ a formula $\neg\varphi$, and, in addition, a computable function which assigns to any φ and $\psi \in L(S)$ a formula $(\varphi \vee \psi)$. (Here $\neg\varphi$, for instance, denotes an $L(S)$-sentence ψ such that $\mathfrak{A} \models_{\mathscr{L}} \psi$ iff not $\mathfrak{A} \models_{\mathscr{L}} \varphi$ (cf. the formulation of Boole($\mathscr{L}$) in §1).

(ii) For every unary U, there is a computable function which associates with every $\varphi \in L(S)$ a formula φ^U.

(iii) There is a computable function which associates with every $\varphi \in L(S)$ a formula $\varphi^r \in L(S^r)$.

$\mathscr{L}_\mathrm{I}$, $\mathscr{L}_\mathrm{II}^\mathrm{w}$, $\mathscr{L}_Q$ are effectively regular logical systems.

Let $\mathscr{L}$ be an effective logical system. We say that $\mathscr{L}$ is *enumerable for validity* if for every decidable S, the set

$$\{\varphi \in L(S) \mid \models_{\mathscr{L}} \varphi\}$$

is enumerable.

Clearly, if $\mathscr{L}$ has an adequate proof calculus then $\mathscr{L}$ is enumerable for validity. (Examples are $\mathscr{L}_\mathrm{I}$ and $\mathscr{L}_Q$.)

Lindström's second theorem tells us that no proper extension $\mathscr{L}$ of $\mathscr{L}_\mathrm{I}$ with LöSko($\mathscr{L}$) can have an adequate proof calculus.

4.4 Lindström's Second Theorem. *Let $\mathscr{L}$ be an effectively regular logical system such that $\mathscr{L}_\mathrm{I} \leq_{\mathrm{eff}} \mathscr{L}$. If* LöSko($\mathscr{L}$) *and if $\mathscr{L}$ is enumerable for validity then $\mathscr{L}_\mathrm{I} \sim_{\mathrm{eff}} \mathscr{L}$.*

PROOF. Let $\mathscr{L}$ satisfy the hypotheses of the theorem. We prove that $\mathscr{L} \leq_{\mathrm{eff}} \mathscr{L}_\mathrm{I}$ in two steps.

First we show

(+) For every decidable S and for every $\psi \in L(S)$ there is a first-order S-sentence φ which has the same models as ψ.

Then we shall prove that the transition from ψ to φ can be carried out effectively: Given a decidable S, we shall set up a *procedure* which yields for every $\psi \in L(S)$ a first-order S-sentence with the same models.

Since $\mathscr{L}$ is an effective logical system, we only need to give a proof of (+) for *finite* decidable S (cf. 4.1). We leave it to the reader to show, using Elim($\mathscr{L}$), that we can furthermore assume S to be relational.

Thus we assume that S is decidable, finite and relational. As a first step we prove

4.5 Lemma. *If for some $\psi \in L(S)$ there is no first-order S-sentence with the same models as ψ, then for every $m \in \mathbb{N}$ there exist S-structures $\mathfrak{A}_m$ and $\mathfrak{B}_m$ such that*

$$(*) \qquad \mathfrak{A}_m \cong_m \mathfrak{B}_m, \qquad \mathfrak{A}_m \models_{\mathscr{L}} \psi, \qquad \mathfrak{B}_m \models_{\mathscr{L}} \neg\psi.$$

Proof. ψ is satisfiable, otherwise $\mathrm{Mod}_{\mathscr{L}}(\psi) = \mathrm{Mod}_{\mathscr{L}_\mathrm{I}}(\exists v_0 \neg v_0 \equiv v_0)$. We proceed indirectly, assuming that for a suitable $m \in \mathbb{N}$ and for all S-structures $\mathfrak{A}$ and $\mathfrak{B}$:

(1) If $\mathfrak{A} \cong_m \mathfrak{B}$, then ($\mathfrak{A} \models_{\mathscr{L}} \psi$ iff $\mathfrak{B} \models_{\mathscr{L}} \psi$).

Let $\varphi_0, \ldots, \varphi_k$ be, up to logical equivalence, the first-order S-sentences of quantifier rank $\leq m$ (cf. XI.3.4). We have (cf. XI.3.7):

(2) $\mathfrak{A} \cong_m \mathfrak{B}$ iff for $i = 0, \ldots, k$, ($\mathfrak{A} \models \varphi_i$ iff $\mathfrak{B} \models \varphi_i$).

For an S-structure $\mathfrak{A}$ let $\varphi_{\mathfrak{A}}$ be the conjunction of the formulas in $\{\varphi_i \mid 0 \leq i \leq k, \mathfrak{A} \models \varphi_i\}$. Then, by (2), for arbitrary $\mathfrak{B}$,

(3) $\mathfrak{A} \cong_m \mathfrak{B}$ iff $\mathfrak{B} \models \varphi_{\mathfrak{A}}$.

Let φ be the disjunction of the (finitely many!) $\varphi_{\mathfrak{A}}$ for which $\mathfrak{A} \models_{\mathscr{L}} \psi$, i.e.,

(4) $\varphi := \bigvee\{\varphi_{\mathfrak{A}} \mid \mathfrak{A}$ S-structure, $\mathfrak{A} \models_{\mathscr{L}} \psi\}$.

We show:

(5) $\mathrm{Mod}_{\mathscr{L}}(\psi) = \mathrm{Mod}_{\mathscr{L}_\mathrm{I}}(\varphi)$,

and thus obtain a contradiction to our assumptions. Suppose first that $\mathfrak{B}$ is a model of ψ. Then $\varphi_{\mathfrak{B}}$ is a member of the disjunction in (4), and since $\mathfrak{B} \models \varphi_{\mathfrak{B}}$, we have $\mathfrak{B} \models \varphi$. Conversely, if $\mathfrak{B} \models \varphi$, there is an $\mathfrak{A}$ such that $\mathfrak{A} \models_{\mathscr{L}} \psi$ and $\mathfrak{B} \models \varphi_{\mathfrak{A}}$ (cf. (4)). Then by (3), $\mathfrak{A} \cong_m \mathfrak{B}$, and finally by (1), $\mathfrak{B} \models_{\mathscr{L}} \psi$. □

To continue with the proof of (+), we assume that there is a sentence ψ in $L(S)$ for which there is no first-order S-sentence with the same models, and aim for a contradiction. By the result 4.5 just proved there are, for every $m \in \mathbb{N}$, S-structures $\mathfrak{A}_m$ and $\mathfrak{B}_m$ such that $\mathfrak{A}_m \cong_m \mathfrak{B}_m$, $\mathfrak{A}_m \models_{\mathscr{L}} \psi$, and $\mathfrak{B}_m \models_{\mathscr{L}} \neg\psi$. Since the assumptions in 3.2 are fulfilled, there is, for some finite S_1, a sentence $\chi_1 \in L(S_1)$ which describes the finite orderings (as explained in 3.2).

We extend S_1 by adding a new unary relation symbol W and consider the $L(S_1 \cup \{W\})$-sentence

$$\vartheta := \chi_1 \wedge \exists x\, Wx \wedge \forall x(Wx \rightarrow x < c)$$

(note that $\mathscr{L}_1 \leq \mathscr{L}$). By the properties of χ_1 (cf. 3.2) we have:

(a) If $\mathfrak{A}$ is an $S_1 \cup \{W\}$-structure such that $\mathfrak{A} \models_{\mathscr{L}} \vartheta$ then W^A is finite and not empty.

(b) For every $m \geq 1$ there is a model $\mathfrak{A}$ of ϑ such that W^A contains exactly m elements.

Thus as $\mathfrak{A}$ ranges over the models of ϑ, W^A ranges over the finite sets (isomorphism property!). We shall now see that we can use (a) and (b) together with Trahtenbrot's theorem to conclude that $\mathscr{L}$ is not enumerable for validity, in contradiction to our assumptions. We argue as in the proof of the incompleteness of second-order logic (cf. X.5.5).

By Trahtenbrot's theorem there is a decidable symbol set S_2 such that the set of fin-valid first-order S_2-sentences is not enumerable. We may assume that S_2 is relational and disjoint from $S_1 \cup \{W\}$.

Let * be a computable function associating with every first-order S_2-sentence φ a sentence $\varphi^* \in L(S_2)$ which has the same models. Then for $\varphi \in L_0^{S_2}$ we have

$$(\circ) \qquad \varphi \text{ is fin-valid iff } \models_{\mathscr{L}} \vartheta \rightarrow (\varphi^*)^W.$$

To prove this, we assume first that φ is fin-valid. If $\mathfrak{A}$ is an $(S_1 \cup \{W\} \cup S_2)$-structure such that $\mathfrak{A} \models_{\mathscr{L}} \vartheta$, then W^A is finite by (a), and thus $[W^A]^{\mathfrak{A}\upharpoonright S_2} \models \varphi$. But then $[W^A]^{\mathfrak{A}\upharpoonright S_2} \models_{\mathscr{L}} \varphi^*$, and hence $\mathfrak{A} \models_{\mathscr{L}} (\varphi^*)^W$. The converse is obtained similarly by applying (b).

Equivalence ($\circ$) enables us to obtain from an enumeration procedure $\mathfrak{P}$ for the set of valid $L(S_1 \cup \{W\} \cup S_2)$-sentences an enumeration procedure $\mathfrak{Q}$ for the fin-valid first-order S_2-sentences, thus yielding a contradiction to Trahtenbrot's theorem. $\mathfrak{Q}$ proceeds as follows for $n = 1, 2, 3, \ldots$: the first (lexicographically) n sentences $\varphi_0, \ldots, \varphi_{n-1}$ from $L_{\mathrm{I}}(S_2)$ are generated, and the $L(S_1 \cup \{W\} \cup S_2)$-sentences $\vartheta \rightarrow (\varphi_0^*)^W, \ldots, \vartheta \rightarrow (\varphi_{n-1}^*)^W$ are formed. (Note that the map * is computable and that the operations of relativization and implication are effective.) Then, using $\mathfrak{P}$, one generates the first n valid $L(S_1 \cup \{W\} \cup S_2)$-sentences, listing those φ_i for which $\vartheta \rightarrow (\varphi_i^*)^W$ occurs. This finishes the proof of $(+)$.

Now, given a decidable S, we describe an effective procedure which associates with every sentence $\psi \in L(S)$ a first-order sentence φ with the same models. Let $\mathfrak{P}$ be an enumeration procedure for the set of valid $L(S)$-sentences, and * a computable function which assigns to every first-order S-sentence χ an $L(S)$-sentence χ^* with the same models.

Given $\psi \in L(S)$, proceed as follows: For $n = 1, 2, 3, \ldots$ use $\mathfrak{P}$ to generate the first n valid sentences $\psi_0, \ldots, \psi_{n-1}$ from $L(S)$; then generate the first (lexicographically) n sentences $\varphi_0, \ldots, \varphi_{n-1}$ from $L_{\mathrm{I}}(S)$, and finally, form the $L(S)$-*sentences* $\psi \leftrightarrow \varphi_0^*, \ldots, \psi \leftrightarrow \varphi_{n-1}^*$. Check whether there are i, j such that $\psi_i = \psi \leftrightarrow \varphi_j^*$. By $(+)$ this must eventually happen. Then let φ_j be the φ associated with ψ. □

Lindström's results initiated a series of investigations of properties of logical systems and relations between them, in a general setting (cf. [2]). In this way it is possible to bring important aspects of such properties into better perspective, thus gaining new insights into concrete logical systems and even into first-order logic. We illustrate this briefly, taking the compactness theorem as an example.

An ordering $(A, <^A)$ which contains no infinite descending chain $\ldots <^A a_2 <^A a_1 <^A a_0$ is said to be a *well-ordering*. All finite orderings are well-orderings, as are $(\mathbb{N}, <^{\mathbb{N}})$ and the ordering which results when $(\mathbb{N}, <^{\mathbb{N}})$ is extended by adding an isomorphic copy. On the other hand, $(\mathbb{Z}, <^{\mathbb{Z}})$ and $(\mathbb{Q}, <^{\mathbb{Q}})$ are not well-orderings.

Let $\mathscr{L}$ be a regular logical system such that $\mathscr{L}_{\mathrm{I}} \le \mathscr{L}$. A well-ordering $(A, <^A)$ is said to be *accessible in* $\mathscr{L}$ (or $\mathscr{L}$*-accessible*) if there is an S with $< \in S$ and an $L(S)$-sentence ψ such that

(a) in every model $\mathfrak{B}$ of ψ, (field $<^B, <^B$) is a well-ordering;
(b) there is a model $\mathfrak{B}$ of ψ such that $(A, <^A) \subset$ (field $<^B, <^B$).

Since $\mathscr{L}_{\mathrm{I}} \le \mathscr{L}$, all finite well-orderings are $\mathscr{L}$-accessible. If $\mathscr{L}$ is compact then no infinite well-ordering is $\mathscr{L}$-accessible. For if a sentence ψ has an infinite model A, where (field $<^A, <^A$) is a well-ordering, then one can show by a method similar to that used in exercise VI.4.11 that ψ has a model $\mathfrak{B}$ in which (field $<^B, <^B$) has an infinite descending chain.

If one assumes LöSko($\mathscr{L}$) and strengthens the regularity conditions slightly, the following two statements are, in fact, equivalent:

(i) not Comp($\mathscr{L}$)
(ii) $(\mathbb{N}, <^{\mathbb{N}})$ is $\mathscr{L}$-accessible.

These considerations motivate us to look beyond the simple dichotomy "Comp($\mathscr{L}$) – not Comp($\mathscr{L}$)", and to make finer distinctions: the more $\mathscr{L}$-accessible well-orderings there are, the more the compactness theorem is violated. As a measure for the violation one can take the smallest well-ordering which is not $\mathscr{L}$-accessible, the so-called *well-ordering number* of $\mathscr{L}$. The study of well-ordering numbers has led to a series of fruitful investigations (cf. [2]). In particular it has turned out that for certain logical systems one can use arguments involving the well-ordering number to compensate for the absence of the compactness property.

References

[1] Barwise, J. *Admissible Sets and Structures*. Berlin-Heidelberg-New York: Springer-Verlag, 1975.

[2] Barwise, J. and Feferman, S. (Editors) *Model-Theoretic Logics*. Berlin-Heidelberg-New York-Tokyo: Springer-Verlag, 1984.

[3] Bolzano, B. *Wissenschaftslehre*, 4 vols. Sulzbach: J. E. von Seidel, 1837.

[4] Chang, C. C. and Keisler, H. J. *Model Theory*. Amsterdam: North-Holland, 1973.

[5] Church, A. A note on the Entscheidungsproblem. *Journal of Symbolic Logic* **1** (1936), 40–41. Correction *ibid.* 101–102.

[6] Cutland, N. *Computability*. Cambridge: Cambridge University Press, 1980.

[7] Dickmann, M. A. *Large Infinitary Languages*. Amsterdam: North-Holland, 1975.

[8] Ebbinghaus, H.-D. *Einführung in die Mengenlehre*. Darmstadt: Wissenschaftliche Buchgesellschaft, 1977.

[9] Enderton, H. B. *Elements of Set Theory*. New York-San Francisco-London: Academic Press, 1977.

[10] Frege, G. *Begriffsschrift, eine der arithmetischen nachgebildete Formelsprache des reinen Denkens*. Halle: Louis Nebert, 1879.

[11] Gödel, K. Die Vollständigkeit der Axiome des logischen Funktionenkalküls. *Monatshefte für Mathematik und Physik* **37** (1930), 349–360.

[12] Gödel, K. Über formal unentscheidbare Sätze der Principia Mathematica und verwandter Systeme I. *Monatshefte für Mathematik und Physik* **38** (1931), 173–198.

[13] Henkin, L. The completeness of the first-order functional calculus. *Journal of Symbolic Logic* **14** (1949), 159–166.

[14] Henle, J. M. and Kleinberg, E. M. *Infinitesimal Calculus*. Cambridge, MA: MIT Press, 1979.

[15] Hermes, H. *Enumerability, Decidability, Computability*. Berlin-Heidelberg-New York: Springer-Verlag, 1969.

[16] Hermes, H. *Introduction to Mathematical Logic*. Berlin-Heidelberg-New York: Springer-Verlag, 1973.

[17] Heyting, A. *Intuitionism*. Amsterdam: North-Holland, 1956.

[18] Hilbert, D. and Bernays, P. *Grundlagen der Mathematik*, I, II. Berlin-Heidelberg: Springer-Verlag, 1934/1939.
[19] Keisler, H. J. Logic with the quantifier "There exist uncountably many". *Annals of Mathematical Logic* **1** (1970), 1–93.
[20] Keisler, H. J. *Model Theory for Infinitary Logic*. Amsterdam: North-Holland, 1971.
[21] Lindström, P. On extensions of elementary logic. *Theoria* **35** (1969), 1–11.
[22] Peano, G. *Arithmetices Principia, Novo Methodo Exposita*. Turin: Fratres Bocca, 1889.
[23] Robinson, A. *Non-Standard Analysis*. Amsterdam: North-Holland, 1966.
[24] Scholz, H. and Hasenjaeger, G. *Grundzüge der mathematischen Logik*. Berlin-Göttingen-Heidelberg: Springer-Verlag, 1961.
[25] Smullyan, R. M. *First-Order Logic*. Berlin-Heidelberg-New York: Springer-Verlag, 1968.
[26] Tarski, A. *Logic, Semantics, Metamathematics. Papers from 1923 to 1938 by A. Tarski*. Oxford: Clarendon Press, 1956.
[27] Tarski, A. Der Wahrheitsbegriff in den formalisierten Sprachen. *Studia Philosophica* **1** (1936), 261–405. English translation in [26].
[28] Tarski, A., Mostowski, A. and Robinson, R. M. *Undecidable Theories*. Amsterdam: North-Holland, 1953.

English translations of parts of [3] and of [10], [11] and [12] can be found in

van Heijenoort, J. *From Frege to Gödel*. Cambridge, MA: Harvard University Press, 1967.

For further reading we refer to the *Handbook of Mathematical Logic*, edited by J. Barwise, Amsterdam: North-Holland, 1977. It contains survey articles on different areas of mathematical logic and also an extensive bibliography.

Index of Notation

Subject Index

Undergraduate Texts in Mathematics

continued from ii

Millman/Parker: Geometry: A Metric Approach with Models.
1981. viii, 355 pages. 259 illus.

Owen: A First Course in the Mathematical Foundations of Thermodynamics
1984. xvii, 178 pages. 52 illus.

Prenowitz/Jantosciak: Join Geometrics: A Theory of Convex Set and Linear Geometry.
1979. xxii, 534 pages. 404 illus.

Priestly: Calculus: An Historical Approach.
1979, xvii, 448 pages. 335 illus.

Protter/Morrey: A First Course in Real Analysis.
1977. xii, 507 pages. 135 illus.

Ross: Elementary Analysis: The Theory of Calculus.
1980. viii, 264 pages. 34 illus.

Sigler: Algebra.
1976. xii, 419 pages. 27 illus.

Simmonds: A Brief on Tensor Analysis.
1982. xi, 92 pages. 28 illus.

Singer/Thorpe: Lecture Notes on Elementary Topology and Geometry.
1976. viii, 232 pages. 109 illus.

Smith: Linear Algebra.
1978. vii, 280 pages. 21 illus.

Smith: Primer of Modern Analysis
1983. xiii, 442 pages. 45 illus.

Thorpe: Elementary Topics in Differential Geometry.
1979. xvii, 253 pages. 126 illus.

Troutman: Variational Calculus with Elementary Convexity.
1983. xiv, 364 pages. 73 illus.

Whyburn/Duda: Dynamic Topology.
1979. xiv, 338 pages. 20 illus.

Wilson: Much Ado About Calculus: A Modern Treatment with Applications Prepared for Use with the Computer.
1979. xvii, 788 pages. 145 illus.